Device Architecture and Materials for Organic Light-Emitting Devices

Sarah Schols

Device Architecture and Materials for Organic Light-Emitting Devices

Targeting High Current Densities and Control of the Triplet Concentration

Springer

Sarah Schols
Large Area Electronics (LAE)
Imec
Kapeldreef 75
3000 Leuven, Vlaams Brabant
Belgium
sarah.schols@imec.be

ISBN 978-94-007-1607-0 e-ISBN 978-94-007-1608-7
DOI 10.1007/978-94-007-1608-7
Springer Dordrecht Heidelberg London New York

Library of Congress Control Number: 2011928673

Cover design: VTeX UAB, Lithuania

Printed on acid-free paper

Springer is part of Springer Science+Business Media (www.springer.com)

Preface

This book is a reprint of the manuscript entitled "Device architecture and materials for organic light-emitting devices", that was submitted to the jury deciding on the achievement of my PhD at the Katholieke Universiteit Leuven in Belgium. It is the result of five years fascinating research at imec in Leuven, Belgium, between 2004 and 2009. Different people contributed to my work and I want to show gratitude to all of them. First, I would like to thank my two promoters, Prof. Paul Heremans (KUL/imec) and Prof. Robert Mertens (KUL/imec) who gave me the opportunity to start a PhD in an inspiring environment as imec. I am also very grateful to the other PhD jury members who took the time to carefully read my manuscript: Prof. Mark van der Auweraer (KUL), Prof. Jan Engelen (KUL), Prof. Henning Sirringhaus (University of Cambridge), Prof. Uli Lemmer (Universität Karlsruhe) and Prof. Adhemar Bultheel (KUL). My special thanks go to Andrey Kadashchuk for introducing me into the exciting world of spectroscopy and for the many scientific discussions, but of course also all other colleagues of the Polymer and Molecular Electronics group at imec with whom I worked together during my PhD are gratefully acknowledged. Besides, I would like to thank Thilo Stöferle (IBM Research Zurich), Tobias Plötzing (RWTH Aachen) and Thorsten Wahlbrink (AMO GmbH) for the nice collaboration. The FWO is acknowledged for the financial support. Finally, I warmly thank my family for their continuous support and interest in my work.

Sarah Schols

Contents

List of Symbols and Abbreviations

α	(*Depending on context*) 1. Loss coefficient (cm^{-1}) 2. Absorption coefficient (cm^{-1})
α_{photon}	Photon loss (cm^{-1})
$A(\lambda)$	Constant related to the cross-section for spontaneous emission
c	(*Depending on context*) 1. Concentration (wt%) 2. Speed of light in free space (299.79×10^6) (m/s)
C_{ox}	Gate dielectric capacitance per unit area (F/m^2)
δ	Density (g/cm^3)
ΔE_T	Difference in triplet energy (eV)
ΔG°	Gibbs free energy (kJ/mol)
Δr	Nuclear displacement (Å)
ΔE_{ST}	Singlet-triplet splitting (eV)
$\varepsilon(\lambda)$	Absorption spectrum
ε_r	Relative dielectric constant
η_{ext}	External quantum efficiency (%)
η_{int}	Internal quantum efficiency (%)
$\eta_{coupling}$	Out-coupling efficiency (%)
E°	Standard cell potential (V)
E_{phot}	Average photon energy (eV)
F	Faraday constant ($\approx$96485) (J/(V mol))
$f(\lambda)$	Photoluminescence spectrum
g	Gain coefficient (cm^{-1})
γ	Charge balance
Γ	Optical confinement factor (%)
h	Planck's constant ($\approx 6.626 \times 10^{-34}$) (Js)
I	(*Depending on context*) 1. Light intensity (W/cm^2) 2. Current (A)
I_{ds}	Drain to source current (A)
I_p	Pumping intensity (J/cm^2)
J	Current density (A/cm^2)

k_{ET}^{F}	Förster energy transfer rate (s^{-1})
k_{ET}^{D}	Dexter energy transfer rate (s^{-1})
K	Constant proportional to the orbital overlap between host and guest
κ	Polarization factor
l	Distance traveled in the gain medium (cm)
L	(*Depending on context*) 1. Sum of the van der Waals radii of host and guest (nm) 2. Transistor channel length (μm)
L_{diff}	Exciton diffusion length (nm)
λ	Wavelength (nm)
λ_{Bragg}	Bragg wavelength (nm)
λ_{exc}	Excitation wavelength (nm)
Λ	Modulation periodicity of a DFB resonator (nm)
m	DFB order number
M	Molecular weight (g/mol)
μ	Mobility (cm^2/Vs)
n	(*Depending on context*) 1. Index of refraction 2. Overall number of electrons exchanged between an oxidizing and reducing agent
n_{eff}	Effective refractive index
N_{exc}	Density of excited states (cm^{-3})
N_{th}	Exciton density at the laser threshold (cm^{-3})
N_A	Avogadro's constant ($\approx 6.022 \times 10^{23}$) ($mol^{-1}$)
ν_i	Vibrational energy level (eV)
p	Pressure (torr)
ϕ_{PL}	Absolute photoluminescence efficiency (%)
P_{meas}	Measured light power (W)
P_{tot}	Total optical power (W)
$\Psi_{molecule}$	Wavefunction of a molecule
$\Psi_{electronic}$	Electronic component of the molecular wavefunction
$\Psi_{nuclear}$	Nuclear component of the molecular wavefunction
q	Elementary charge ($\approx 1.602 \times 10^{-19}$) (C)
Q	Resonator quality factor
R	Mean distance between host and guest (nm)
R_0	Effective Förster radius (nm)
r_{st}	Singlet/triplet ratio of excitons
ρ	Average distance between dopant molecules (nm)
S_i	Singlet exciton energy level (eV)
S	Subthreshold slope (V/dec)
σ_{RZ}	Exciton density per unit area in the recombination zone (cm^{-2})
$\sigma_{SE}(\lambda)$	Cross-section for stimulated emission (cm^2)
σ_{TT}	Cross-section for triplet-triplet absorption (cm^2)
T_i	Triplet exciton energy level (eV)

T_{bake}	Baking temperature (°C)
t	Thickness of the dielectric layer (nm)
t_{del}	Time delay (s)
τ	Lifetime (s)
τ_r	Radiative lifetime (s)
V_{ds}	Drain to source voltage (V)
V_{gs}	Gate to source voltage (V)
V_{on}	Onset voltage (V)
V_{th}	Threshold voltage (V)
W	Transistor channel width (μm)
AFM	Atomic force microscopy
Ag	Silver
Al	Aluminum
Alq_3	Tris-(8-hydroxyquinoline) aluminum
ASE	Amplified spontaneous emission
AZO	Aluminum doped zinc oxide
BARC	Bottom anti reflection coating
BCl_3	Boron trichloride
BSP-Me	1,4-Bis(4-methylstyryl)benzene
$Btp_2Ir(acac)$	Bis(2-(2'benzothienyl)pyridinato-N,$C^{3'}$)(acetylacetonate)-iridium(III)
CAMFR	Cavity modelling framework
CCD	Charge coupled device
CH_2	Methylene
CNPPP	2-[(6-cyano-6-methylheptyloxy)-1,4-phenylene] copolymer
COT	1,3,5,7-cyclooctatetraene
Cr	Chromium
CW	Continuous-wave
C_4F_8	Octafluorocyclobutane
DBR	distributed Bragg reflector
DCM	4-(dicyanomethylene)-2-methyl-6-[(4-dimethylaninostyryl)-4H-pyran
DCM_2	4-(dicyanomethylene)-2-methyl-6-(julolindin-4-yl-vinyl)-4H-pyran
DF	Delayed fluorescence
DFB	Distributed feedback
DFH-4T	α,ω-diperfluorohexyl-quaterthiophene
DFHCO-4T	5,5'''-diperfluorohexylcarbonyl-2,2':5',2'':5'',2'''-quaterthiophene
2D	Two-dimensional
DOS	Density of states
EBL	Electron beam lithography
EL	Electroluminescence
ETL	Electron-transporting layer
Fe	Iron

FWHM	Full-width-half-maximum
$F_{16}CuPc$	Copper hexadecafluorophthalocyanine
He	Helium
HOMO	Highest occupied molecular orbital
HSQ	Hydrogen silsesquioxane
HTL	Hole-transporting layer
H_2O	Water
IC	Internal conversion
ICP-RIE	Inductively coupled plasma reactive ion etching
IL	Interference lithography
ISC	Intersystem crossing
ITO	Indium tin oxide
LEOFET	Light-emitting organic field-effect transistor
LiF	Lithium fluoride
LPPP	Ladder-type poly(para-phenylene)
LUMO	Lowest unoccupied molecular orbital
MEH-PPV	Poly(2-methoxy-5-(2'-ethyl-hexyloxy)-1,4-phenylene-vinylene)
MeLPPP	Methyl-substituted ladder-type poly(para-phenylene)
Mg	Magnesium
MTR	Multiple trapping and release model
N_2	Nitrogen
OLED	Organic light-emitting diode
OMA	Optical multichannel analyzer
OMBD	Organic molecular beam deposition
OTFT	Organic thin film transistor
O_2	Oxygen
PαMS	Poly-α-methylstyrene
PF	Prompt fluorescence
PF2/6	Poly(9,9-di(ethylhexyl)fluorene)
Ph	Phosphorescence
PL	Photoluminescence
PMMA	Poly(methyl metacrylate)
PmPV-co-DOctOPV	Poly(*m*-phenylenevinylene-*co*-2,5-dioctoxy-p-phenylenevinylene)
PPV	Poly(phenylene vinylene)
PS	Polystyrene
PTAA	Poly(triarylamine)
PTCDA	3,4,9,10 perylenetetracarboxylic dianhydride
PTCDI-$C_{13}H_{27}$	N,N'-ditridecylperylene-3,4,9,10-tetracarboxylic diimide
SEM	Scanning electron microscopy
SEP	Standard electrode potential
SF	Superfluorescence
SHG	Second harmonic generation
SiO_2	Silicon dioxide

SMU	Source-measure unit
spiro-SBCz	2,7-bis[4-(N-carbazole)phenylvinyl]-9,9'-spirobifluorene
Ta_2O_5	Tantalum pentoxide
TCO	Transparent conductive oxides
TE	Transverse electric
THF	Tetrahydrofuran
Ti	Titanium
TiO_2	Titanium dioxide
TMAH	Tetramethylammonium hydroxide
TSL	Thermally stimulated luminescence
UV	Ultra violet
VR	Vibronic relaxation
Yb	Ytterbium
Zn	Zinc

Chapter 1
Introduction

This first chapter provides an introduction to the basic principles of optoelectronics, which will be useful for the design and operational understanding of the optoelectronic device and material combinations presented in this book. It starts with a description of the general properties of organic semiconductors. We focus on the principle of conjugation and briefly discuss transport in disordered organic semiconductors. Attention is also given to the different radiative and nonradiative transitions possible in organic compounds. Organic light-emitting diodes and organic light-emitting transistors are the subject of the second section. The working principle of both devices is discussed and contrasted. Section 1.3 then focuses on another optoelectronic device, the organic semiconductor laser. First, general aspects of laser activity and the motivation for plastic lasers are considered. Next, lasing in organic semiconductors and the difficulties in achieving electrical pumping are discussed. As a conclusion to this introduction, Sect. 1.4 gives an outline of the subsequent chapters of this book.

1.1 Organic Semiconductors

In general, organic materials are compounds with a sequence of carbon atoms as backbone. Originally these carbon-based materials were considered as being insulating. In the 1950s, photoconductivity was reported for the first time in organic crystals such as anthracene and naphthalene [1–3]. The main breakthrough for organic electronics, however, took place in 1977 when Shirakawa, MacDiarmid and Heeger demonstrated that the conductivity of conjugated polymers could be varied over the full range from insulator to metal by chemical doping [4]. Since then, remarkable progress has been made in synthesizing conjugated polymers and organic small molecules, in understanding their properties, and in developing them for use in electronic and optical devices.

S. Schols, *Device Architecture and Materials for Organic Light-Emitting Devices*,
DOI 10.1007/978-94-007-1608-7_1,

1.1.1 Semiconducting Properties of Organic Materials

The semiconducting characteristics of several organic materials are the result of the properties of the atomic carbon atom. The electronic configuration of the ground state of a carbon atom is $1s^2 2s^2 2p^2$. In this configuration only the two unpaired p electrons in the outer electronic shell are able to form a bond with another atom, leading to e.g. the formation of methylene (CH_2). Mostly, however, the bonding is described by hybridization. Hybridization of the different atomic orbitals leads to more stable molecules and makes it possible to obtain a new set of hybrid orbitals, having an orientation in space matching the actual geometry of the compound much better [5]. Different hybridizations are possible: sp^3, sp^2 and sp hybridization (Fig. 1.1).

For organic semiconducting materials, the sp^2 hybridization is the most important one. In this case, the 2s orbital and two of the three 2p orbitals will combine to form three sp^2 hybrid orbitals. These three sp^2 hybrid orbitals are coplanar, having an angle of 120° between them (Fig. 1.1(b)), and will form three covalent σ-bonds with neighboring atoms, which are strongly localized in molecular σ-orbitals. The remaining 2p orbital, which is not used during hybridization, is positioned perpendicular to this plane. If two unhybridized 2p orbitals of neighboring atoms overlap, an additional π-bond is formed via π-orbitals. The overlap region and thus the electron density of this π-bond is located above and below the plane of the σ-bonds. The combination of a sp^2-sp^2 σ-bond and a 2p-2p π-bond between two neighboring carbon atoms is referred to as a double bond (C=C). If the number of sp^2 hybridized carbon atoms is increased, a conjugated system with alternating single and double bonds is formed. The simplest example of such a system is polyacetylene. In conjugated systems, the π-orbitals ideally extend over the complete molecule (the π-molecular orbital) and a delocalized electron cloud is created. The delocalized electrons are shared by all atoms of the conjugated molecule and can move freely over the entire molecule.

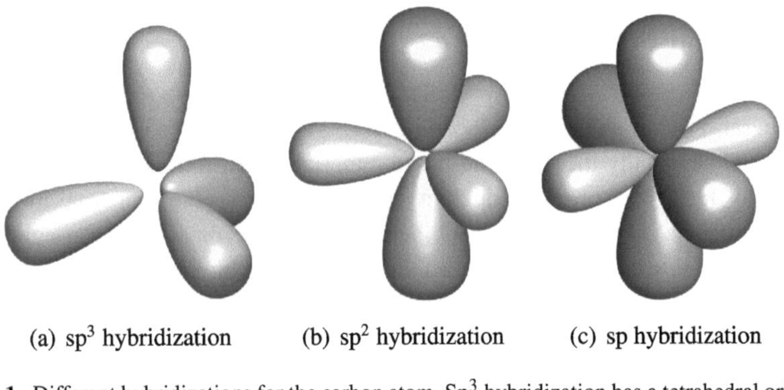

(a) sp^3 hybridization (b) sp^2 hybridization (c) sp hybridization

Fig. 1.1 Different hybridizations for the carbon atom. Sp^3 hybridization has a tetrahedral oriented structure, sp^2 hybridization forms a trigonal-planar geometry and the two sp hybrid orbitals are oriented along the same axis

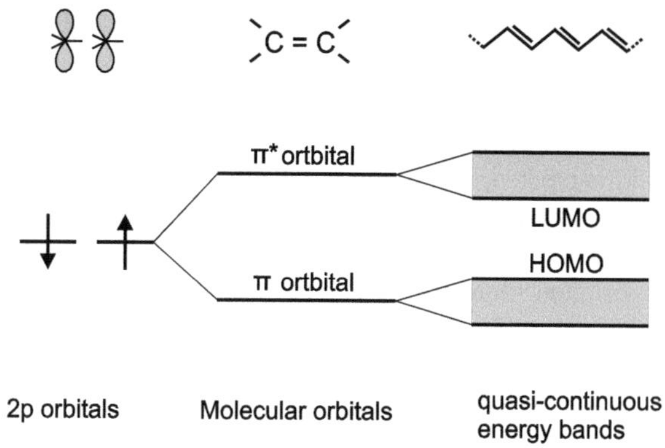

Fig. 1.2 Schematical representation of the splitting of two 2p orbitals into a bonding molecular π-orbital and an anti-bonding molecular π^*-orbital. Increasing the number of carbon atoms, leads to more degeneration and the formation of quasi-continuous bands of occupied and unoccupied states

In general, overlap and (linear) combination of orbitals give rise to the formation of new, more extended orbitals. This process is denoted by the term 'splitting'. The energy of these new orbitals with respect to each other and the original orbitals depends on the energy and overlap of the original orbitals: the stronger the overlap, the larger the splitting. In Fig. 1.2 we combine the 2p orbitals of two carbon atoms to a bonding molecular π-orbital and an anti-bonding molecular π^*-orbital. As can be seen from Fig. 1.2 the bonding molecular π-orbital and the anti-bonding molecular π^*-orbital are, respectively, lower and higher in energy than the original atomic orbitals. Every 2p orbital has only one electron and therefore, according to the Pauli principle, only the bonding molecular π-orbital will be occupied in the ground state. The anti-bonding π^*-orbital is unoccupied. When more atoms are added, more bonding and anti-bonding orbitals with different energies are created. Clearly, if the number of carbon atoms goes to infinity, the π and π^* energy levels will form quasi-continuous bands. The highest molecular orbital that still contains electrons at 0 K is called the highest occupied molecular orbital (HOMO). The lowest molecular orbital that has no electrons at 0 K is referred to as the lowest unoccupied molecular orbital (LUMO). Typically, the HOMO and LUMO are separated by an energy gap, which induces semiconducting properties in these materials. The larger the molecule, the broader the energy bands and thus the smaller the band gap will be. In general, the energy gap in organic semiconducting materials is relatively large (2 to 5 eV) compared to inorganic semiconductors (1.1 eV for silicon and 1.4 eV for GaAs [6]).

Usually, in organic electronics thin solid films are studied rather than single organic molecules. Similar to the molecular level, molecular orbitals of neighboring conjugated molecules will interact and overlap. According to molecular orbital theory these molecular orbitals will split again and form intermolecular orbitals. The intermolecular overlap, however, is typically much smaller than the intramolecular overlap because of the larger distances and weaker interactions between molecules than those between atoms. Therefore, splitting and intermolecular delocalization are significantly smaller and electrons will have much less tendency to be delocalized over the complete film. As a consequence, charge transport is not limited by the transport within one molecule, but rather by the transport from one molecule to another. This clearly indicates that in order to obtain efficient charge transport the intermolecular overlap has to be optimized.

So far, the energy states were assumed to be located in a neutral crystal without excess charge. However, when an excess charge is placed on a molecule, an electric field is created, polarizing the environment in order to stabilize the excess charge. As a consequence, the molecule and its nearest neighbor environment relax their structure to obtain a new equilibrium. Different polarization effects can take place at different time scales: electronic polarization, molecular polarization and lattice polarization [7]. These polarization effects significantly lower the energy gap [7] and strongly influence charge transport. The charge together with its accompanying polarization forms a quasi-particle that is referred to as a polaron. Generally, hopping transport is seen for organic materials, where polarons jump from one molecule to the next [8]. The structural relaxation during polaron formation also results in the creation of two new energy levels within the band gap, allowing new electronic transitions (e.g. polaron absorption).

1.1.2 Charge Transport in Organic Materials

Similar to inorganic semiconductors, charge transport in organic materials occurs via drift and diffusion. The exact transport mechanism, however, is much more complicated due to the more complex, molecular nature of organic materials. As mentioned before, charge transport through the material is impeded due to polarization effects and due to the larger intermolecular distances and smaller intermolecular orbital overlap compared to inorganic semiconductors. Several theories have been proposed to describe charge transport in organic semiconducting materials, however, none of them can explain all experimental observations. Probably, several mechanisms occur at the same time and depending on the specific conditions charge transport is dominated by one of these mechanisms. In this section a very brief overview of the models that are often used to explain charge transport in organic semiconductors is given. For a more complete overview of relevant charge transport theories we refer to literature [7, 9, 10].

Band Transport Model

In the band model intermolecular interactions lead to the formation of energy bands in which electrons and holes can be transported freely. Band transport thus occurs in delocalized states and is limited by scattering of lattice vibrations. Since these lattice vibrations are reduced at low temperatures, the band transport model suggests that the charge carrier mobility increases with decreasing temperature. Typically, band transport is observed in inorganic semiconductors, which are highly ordered with strong coupling between the atoms. Most organic semiconductors, however, are characterized by a high degree of disorder and the weak electronic coupling (Van der Waals and dipole-dipole interactions) between different molecules can easily be broken, causing localized instead of delocalized states. Consequently, band transport is generally not the transport mechanism in organic semiconductors. Only for highly ordered molecular crystals such as naphtalene [11], anthracene [12], rubrene [13] and pentacene [14], this behavior, indicated by an increase of the mobility at lower temperatures, has been observed.

Multiple Trapping and Release Model

The multiple trapping and release model (MTR) is originally developed to describe charge transport in hydrogenated amorphous silicon [15], but it is sometimes used to explain transport in disordered organic materials [16]. The model assumes that charge transport occurs through delocalized, extended states. Transport is, however, impeded by impurities, defects, grain boundaries, etc., which create a distribution of traps near the transport band. During charge transport charge carriers can be trapped. These trapped carriers can then, after a certain time, be thermally released to reach the transport band, where they can be trapped again. The time the carriers are trapped depends strongly on the temperature and on the depth of the trap. The higher the temperature and the shallower the trap, the faster the carriers are released. The MTR model predicts an Arrhenius-like temperature dependence of the mobility [15]. In organic semiconductor films the distribution of energy levels below the nominal LUMO or HOMO is to a large extent due to disorder. The MTR model has been used quite successfully to describe transport in these films [16, 17], despite the fact that the—more complex—hopping model, described below, is physically more relevant.

Hopping Transport Model

Because of the weak intermolecular coupling in disordered organic semiconductors, the states for charge carriers in these material systems are considered as localized instead of delocalized. Charge carrier transport is then typically described by hopping [18, 19], i.e., thermally activated tunneling of carriers between localized states. In the hopping transport model, the mobility of the charge carriers depends on their energy within the density of states distribution and increases if the density of neighboring states (in space) is large, and/or if there are states available at lower energy [7]. As mentioned before, charge transport in organic disordered semiconductors is mainly influenced by polarization effects. Polarization acts as a potential well, impeding the movements of the charge. Hopping can then be seen as polarons jumping from one site to the next. At elevated temperatures, thermal energy is sufficient to overcome small energy barriers. At low temperature, however, transport only takes place via tunneling. In general, charge transport in disordered organic semiconductors is described as a series of carrier hops from one site to the next, followed by polaronic relaxation.

1.1.3 Optical Transitions in Organic Materials

A molecular system typically consists of various energy levels. These energy levels are mainly electronic, but also vibrational, rotational and translational levels exist. Upon optical excitation of a molecule in the UV/VIS spectral range an electron can be promoted from the HOMO to the LUMO or to a higher located empty orbital. Similar to the presence of excess charges, an electric field is created and the surrounding molecules are polarized. The quasi-particle of an excited molecule that is formed in this way is an excitonic polaron, but more often the term 'exciton' is used.

The different energy levels of a molecule and the transitions between them can be described by a Jablonski diagram [20], as shown in Fig. 1.3. In this diagram the states are arranged vertically by energy and grouped horizontally by spin multiplicity. A distinction is made between singlet excitons and triplet excitons. A singlet exciton is characterized by the presence of two unpaired electrons with opposite spin (total spin-momentum is equal to 0), whereas a triplet exciton comprises two unpaired electrons with equal spin (total spin-momentum equals 1). Because of the repulsive nature of the spin-spin interactions between electrons of the same spin, a triplet state has a lower energy than the corresponding singlet state. In the diagram the singlet and triplet states are arranged and numbered in order of increasing energy (S_i and T_i). S_0 represents the ground state, which is intrinsically a singlet state in almost all organic compounds. Each electronic level possesses a vibrational manifold with typical sublevel spacing on the order of $\sim$0.1 eV. Each vibronic sublevel corresponds for his part to a manifold of rotational and translational levels with $\sim$0.01 eV internal spacing. In general, radiative transitions are indicated by straight arrows and nonradiative transitions by squiggly arrows. The vibrational ground states of each electronic state are indicated with thick lines, the higher vibrational states with thinner lines.

Absorption of a photon can promote an electron from the ground state to an excited state. Such an excited state can release its excess energy via different mechanisms. A first way is by the emission of a photon, which is called fluorescence or phosphorescence depending on the multiplicity of the excited state. Fluorescence is a rapid radiative process in which the spin multiplicities of the initial and the final states are the same. Phosphorescence, on the other hand, involves the transition from the T_1 excited state to the S_0 ground state, requiring spin-orbit coupling to conserve total momentum. Because this transition is spin-forbidden, its intensity is orders of magnitude smaller than the intensity of fluorescence from singlet excited states. For the same reason the lifetime of the triplet state is very long, up to ms or even sec, compared to ns for fluorescence [20]. During this long lifetime of the triplet state, the exciton is liable to diffusion towards trap states or to bimolecular reactions with other excitations such as triplet excitons, singlet excitons or charge carriers. These competing processes often prevent the observation of phosphorescence, especially at room temperature. Apart from radiative transitions, the excited state can also release its excess energy via nonradiative processes. Relaxation of the excited state to its lowest vibrational level is called vibronic relaxation (VR). This

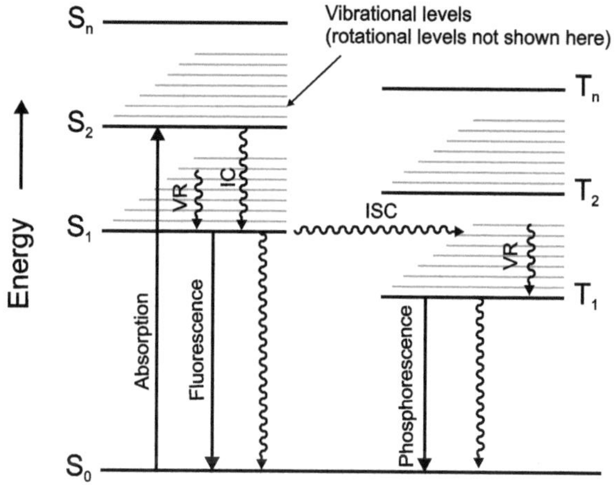

Fig. 1.3 Jablonski diagram illustrating the different energy levels of an organic molecule and indicating the possible transitions between them

process involves the dissipation of energy from the molecule to its surrounding environment and is very fast ($<10^{-12}$ s) [21]. VR is enhanced by physical contact of an excited molecule with other particles with which energy, in the form of vibrations and rotations, can be transferred through collisions. A second type of nonradiative transitions is internal conversion (IC), which occurs when a vibrational state of an electronically excited state couples to an isoenergetic vibrational state of a lower electronic state with the same spin multiplicity. IC between higher excited states ($S_n \rightarrow S_m$ and $T_n \rightarrow T_m$) occurs on the sub ps-timescale [22]. On the other hand, IC from the first excited singlet state S_1 is much slower, making fluorescence possible. A third type is intersystem crossing (ISC); this is a radiationless transition to a state with a different spin multiplicity. In molecules with large spin-orbit coupling, intersystem crossing is much more important than in molecules that exhibit only small spin-orbit coupling. Usually the triplet state is populated via ISC from the optically excited singlet state.

A very important approximation to describe the physics of an excited molecule is the Born-Oppenheimer approximation. This approximation states that the motion of the electrons can be decoupled from the motion of the nuclei because electrons have less mass and therefore respond approximately instantaneously to the movement of the nuclei. The Born-Oppenheimer approximation makes it possible to compute the wavefunction $\Psi_{molecule}$ of a molecule in two consecutive steps; it allows the wavefunction to be broken into its electronic $\Psi_{electronic}$ and nuclear (vibrational, rotational) $\Psi_{nuclear}$ components:

$$\Psi_{molecule} = \Psi_{electronic} \times \Psi_{nuclear}. \tag{1.1}$$

In other words, the energy levels of a molecule may be determined by calculating the electron energies for a fixed nuclear position. By repeating this calculation for different nuclear arrangements, the equilibrium structure of the molecule can be determined and the molecular potential energy curve as function of the configurational coordinates can be constructed (Fig. 1.4). In such a diagram an optical transition is represented by a vertical line according to the Franck-Condon principle [20]. This

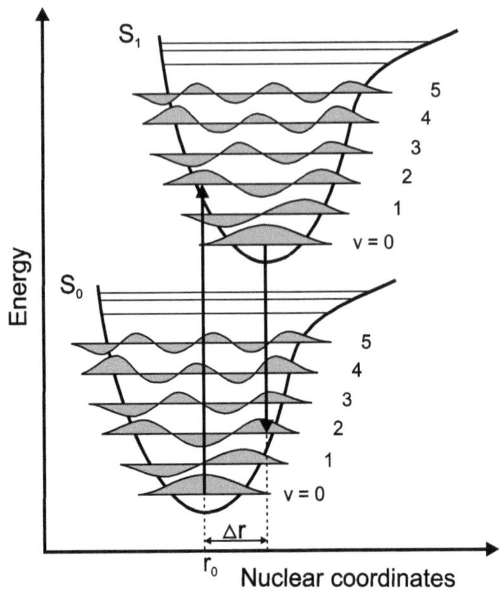

Fig. 1.4 Molecular potential energy curves for the ground state S_0 and the excited state S_1 of a diatomic molecule. According to the Franck-Condon principle, electronic transitions are very fast compared with nuclear motions and therefore vibrational levels are favored when they correspond to a minimal change in the nuclear coordinates. In this particular case transitions between $\nu = 0$ and $\nu = 2$ will be favored. r_0 is the equilibrium distance and Δr is the nuclear displacement. Absorption and emission are indicated by the vertical arrows

principle, which is the analog of the Born-Oppenheimer approximation for optical transitions, states that electronic transitions are essentially instantaneous compared with the time scale of nuclear motions. Therefore if the molecule has to move to a new vibrational level during the electronic transition, this new vibrational level must be instantaneously compatible with the nuclear position and momentum of the vibrational level of the molecule in the originating electronic state. After electronic transition the molecule will relax and obtain its equilibrium configuration. Figure 1.4 represents the potential energy curves for the ground state S_0 and the excited state S_1 of a diatomic molecule. Upon absorption of a photon of the necessary energy, an electronic transition from the $\nu = 0$ vibrational level of the ground state to the excited state occurs, indicated by a vertical line connecting the two potential energy surfaces. In the figure the situation where excitation leads to a geometrical change is illustrated. This geometrical change is indicated by a shift in nuclear coordinates between the S_0 and the S_1 state (Δr). According to the Franck-Condon principle, the probability that the molecule can end up in any particular vibrational level is proportional to the square of the (vertical) overlap of the vibrational wavefunctions of the original and final state. For this particular example, this means that the $(0 \rightarrow 2)$ vibrational transition will be the strongest. After electronic transition the molecule quickly relaxes to the lowest vibrational level, and from there it can decay to the S_0 state via photon emission. The Franck-Condon principle is applied equally to absorption and to fluorescence. As a consequence, the emission spectrum is red-shifted with respect to the absorption spectrum. This red-shift is known as the Stokes shift and varies form 0.1 eV to 1 eV in most organic compounds [21].

1.2 State of the Art Organic Light-Emitting Devices

Since the first report of electroluminescence (EL) in organic thin films [23], organic light-emitting devices have gained significant attention. In the last decade remarkable progress has been achieved in the development of organic light-emitting diodes using polymers as well as small molecule semiconductors [24]. Today, they are considered to be one of most promising technologies for display and lighting applications. Another class of organic light-emitting devices that is increasingly attracting interest is the organic light-emitting transistor. The main aspects of both device architectures will be discussed in this section.

1.2.1 Organic Light-Emitting Diodes

Although the first EL from organic materials was reported in 1953 [25, 26], it was only in 1987 that the first efficient organic light-emitting diode (OLED) comprising a bilayer heterojunction structure was demonstrated [23]. For the first time a reasonable brightness was achieved from an organic light-emitting device when a voltage below 10 V was applied. Shortly afterwards, in 1990, green-yellow EL was obtained from the conjugated polymer PPV[1] in a single layer device structure [27]. Since then, there has been an increasing research activity in the field of organic electroluminescent devices. Considerable effort has been put into the design of new materials and the improvement of color purity, luminescence efficiency and device stability [24, 28]. Today, one has succeeded in solving most critical issues and the first OLED applications are already commercially available. In the future, OLEDs may play an important role in flat panel displays and ambient lighting [29].

OLEDs are typically fabricated by sandwiching one or more appropriate organic layers between two conductive electrodes, a concept invented by S.A. VanSlyke and C.W. Tang [30]. To allow the emitted light to emerge from the device, one of the electrodes is made transparent. The conductive oxide indium tin oxide (ITO) is commonly used for this purpose. When a positive bias is applied to the ITO, electrons and holes are injected into the device and under influence of the applied electrical field these carriers travel through the organic layer or organic layers. It is shown that the current in OLEDs is typically space charge limited [9, 31], i.e., limited by the bulk of the semiconductor. In other words, the measured current is a drift current, determined by the mobility of the charge carriers. Charge transport occurs in the vertical direction, perpendicular to the stack of the organic layers, and the carrier mobility in the organic semiconductor element of OLEDs is typically low [31, 32]. When a pair of oppositely charged carriers meet, excitons are formed and radiative relaxation of these excitons results in light-emission. A schematic representation of the working principle of an OLED is illustrated in Fig. 1.5.

[1] Poly(phenylene vinylene).

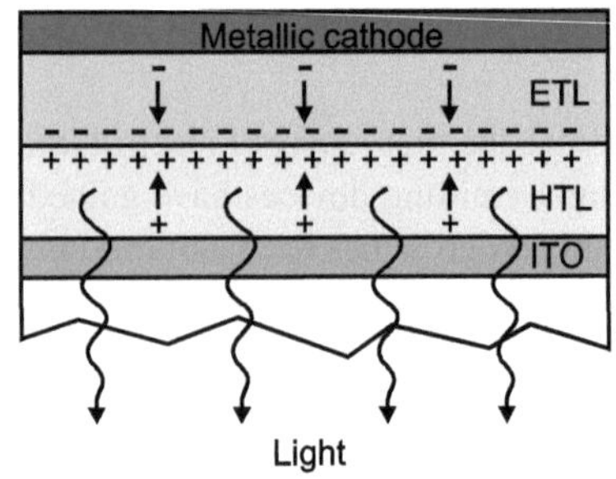

Fig. 1.5 Schematic illustration of the working principle of a basic two-layer OLED, comprising a hole-transporting layer (HTL) and an electron-transporting layer (ETL)

The simple structure, shown in Fig. 1.5 can be modified to a three-layer architecture, in which an additional luminescent layer is introduced between the hole-transporting layer (HTL) and the electron-transporting layer (ETL). The light-emitting layer can then be chosen to have a high luminescence efficiency, whereas the ETL and the HTL can be optimized for efficient charge transport. Such hetero-junction concepts have been widely used in high-performance OLEDs [33, 34].

The external EL quantum efficiency η_{ext} in OLEDs is limited by the charge balance γ, the out-coupling efficiency $\eta_{coupling}$, the photoluminescent efficiency of the material ϕ_{PL} and the singlet/triplet ratio of excitons r_{st} formed by electrical injection [35]:

$$\eta_{ext} = \gamma r_{st} \phi_{PL} \eta_{coupling} = \eta_{int} \eta_{coupling}. \tag{1.2}$$

On average one singlet and three triplets are created for each four electron-hole pairs injected in the active organic semiconductor layer of an OLED [36]. However, as discussed in Sect. 1.1.3 only singlet excitons can decay efficiently and emit light. The radiative relaxation of triplets is a forbidden relaxation process and therefore triplet excitons normally do not contribute to the luminescence of OLEDs. This limits the maximum internal quantum efficiency η_{int} to only 25% when fluorescent light-emitting molecules are used. Taking into account an out-coupling efficiency of ~20%, this means that the maximum η_{ext} of a fluorescent OLED is limited to ~5%. η_{int}, however, can be increased by incorporating heavy atom organometallic compounds as guest dopants in the light-emitting layer. This increases spin-orbit coupling, and thus enhances the spin-flip process necessary for radiative decay of triplet excitations. The incorporation of platinum and iridium-containing organometallic compounds in different hosts to harvest light-emission from triplet states in OLEDs was successfully demonstrated by Baldo *et al.* in 1998 [37]. Devices based on these phosphorescent dyes represent the best performing OLEDs to date, with η_{int} approaching 100% [34] and η_{ext} between 20% and 30% [38, 39].

The quantum efficiency of OLEDs is further limited by the photoluminescent efficiency ϕ_{PL} of the organic light-emitting material (Eq. 1.2). The photoluminescent efficiency of a molecule under optical excitation is defined as the number of emitted photons per absorbed photon and is usually measured by placing the sample in an integrating sphere. Considerable effort has gone into increasing ϕ_{PL} by optimizing the chemical design of the emissive material [40] and by incorporation of the light-emitting dye in a host matrix [41]. The third limiting factor indicated by Eq. 1.2, the charge carrier balance γ, can be improved by reducing the injection barriers at the electrode/organic interface [42] and by matching the mobility of the holes and electrons [35], whereas roughening the substrate [43] and the use of low refractive index substrates and transparent contacts [44] may increase $\eta_{coupling}$ of OLEDs.

1.2.2 Organic Light-Emitting Transistors

Complementary to the traditional vertical light-emitting diodes, light-emitting organic field-effect transistors (LEOFETs) have been proposed as lateral light-emitting devices. Because they combine the optical output of an OLED and the gate control of an organic field-effect transistor in one single device, they may become interesting structures in the field of organic displays. Displays based on LEOFETs may eliminate the difficult integration of an organic light-emitting structure and the organic driving backplane. In addition, LEOFETs can be used to study the opto-electronic performance of organic semiconductors. In contrast to OLEDs, where charge transport occurs perpendicularly to the organic layers, charge transport in LEOFETs occurs in the plane and the carriers are transported by field-effect [45]. In field-effect transistors the current flowing between the source and the drain is modulated by applying a bias to a third contact, the gate electrode. In this way, charge carriers can be accumulated or depleted in the semiconductor close to the semiconductor/insulator interface. The gate field thus provides an additional control for the amount of charges in the organic semiconductor. In addition, the accumulated charges can flood deep traps, so that the effective mobility of the remaining carriers is larger. Typically, field-effect mobility is several orders of magnitude higher than the charge carrier mobility in a conventional OLED [46].

Although the basic concept of a LEOFET dates from 1996 [47], the development of LEOFETs is still in a relatively embryonic stage. The working principle of a LEOFET is based on simultaneous injection of electrons and holes in a double layer or an ambipolar layer by adjusting the gate-source voltage and the drain-source voltage. When the biasing conditions are so that the potential at some point in the channel equals the gate potential, the accumulated charge at that point is zero and, consequently, the electron and hole accumulation layers vanish there. Exciton formation occurs near to that point and radiative relaxation of these excitons to the ground state leads to light emission. This working principle is schematically illustrated in Fig. 1.6.

The first LEOFET was demonstrated by Hepp *et al.* [48] and was based on vacuum evaporated tetracene as the organic semiconductor. Because of a large energy barrier between the electron-injecting electrode and the LUMO of tetracene, only few electrons could be injected in the organic semiconductor. In addition, electron transport is negligible in tetracene [49], resulting in exciton formation and recombination in the immediate neighborhood of the electrode. Since 2003, many other results on tetracene based LEOFETs have been reported [50–52]. Similar LEOFETs,

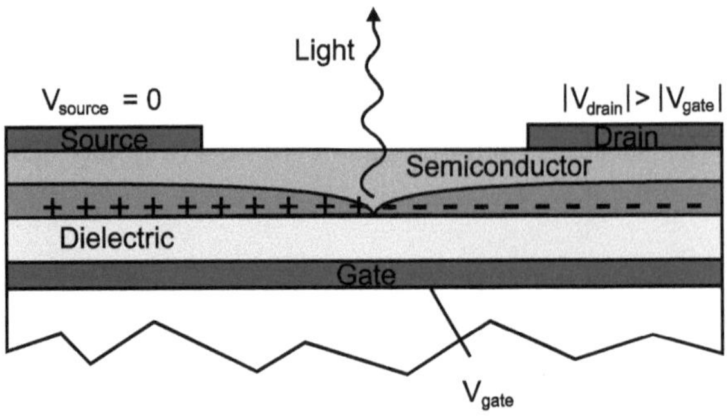

Fig. 1.6 Schematic illustration of the working principle of an ambipolar LEOFET under balanced electron and hole injection. Light-emission occurs where the electron and hole accumulation layers vanish

based on one unipolar organic semiconductor, have been realized by using polymers [53–56] as well as doped small molecules [57] as the organic semiconductor. Electroluminescence in all of these unipolar LEOFETs occurs very close to the electron-injecting metal electrode, which is one of the main drawbacks of this approach. As an alternative strategy to enhance device performance, i.e. reducing exciton quenching and light absorption at the metallic electrodes, ambipolar LEOFETs were proposed. Ambipolar LEOFETs in which the active layer is formed by a heterostructure of a p-type and an n-type organic semiconductor, either by co-evaporation [58, 59] or by subsequent evaporation of both semiconductors [60–62], were reported by different groups. Very recently, Namdas *et al.* demonstrated a solution processed bilayer LEOFET [63]. Due to the presence of an n-type and p-type organic semiconductor, electron transport occurs in a more favorable way, compared to p-type-only LEOFETs. Therefore, light-emission does not necessarily happen near the metal electrode in these devices. Another LEOFET architecture, which is also based on a heterostructure of a p-type material and en n-type material, is proposed by De Vusser *et al.* [64]. This LEOFET makes it possible to fix the location and the dimensions of the pn-junction with photolithographical resolution. In this way light-emission can be obtained inside the transistor channel, far away from the metallic electrodes. LEOFETs based on a single polymeric [65–69] or small molecule [70] ambipolar semiconductor were also reported. Zaumseil *et al.* demonstrated that the light-emission zone could be spatially controlled by varying the bias condition in such an ambipolar LEOFET [65]. Depending on the biasing conditions, the recombination zone can be moved inside the transistor channel. A near-infrared ambipolar LEOFET [71] and analytic device models for the device operation [72] and the recombination profile [73] of ambipolar LEOFETs have been proposed by Smits *et al.* More recently, light-emitting transistors based on tetracene [74], rubrene [75] and BSP-Me[2] [76] single crystal have been reported. For a detailed overview on recent progress in the field of LEOFETs, we refer to several review papers that can be found in literature [49, 77, 78].

[2] 1,4-Bis(4-methylstyryl)benzene.

1.3 Organic Semiconductor Lasers

Apart from OLEDs and LEOFETs, other optoelectronic devices have also been demonstrated in literature [79–83]. Recently, organic semiconducting materials were highlighted as a new class of solid-state laser materials [84–86]. In this section, lasing in organic semiconductors is discussed and the challenges for achieving electrically pumped lasing are put forward.

1.3.1 General Aspects of Laser Action

The first laser was demonstrated on 16 May 1960 by Theodore Maiman at Hughes Research Laboratories, by shining a high-power flash lamp on a ruby rod with silver-coated surfaces [87]. Since then, lasers have become an important instrument, not only in physical research but in almost each field of everyday life. Optical storage devices such as compact disc and DVD players are by far the largest single application of lasers, but also bar code readers, laser printers and laser pointers are well-known examples of products where laser emission is used. Other common application fields for lasers are fiber-optic communication and medicine.

The term LASER is an acronym for 'Light amplification by stimulated emission of radiation' [88] and suggests that a laser is a device emitting light through the process of stimulated emission. However, this is also true for related phenomena such as amplified spontaneous emission (ASE) or superfluorescence (SF) (Sect. 1.3.3). Therefore, seen from a laser technological point of view, a definition of the term lasing derived directly from the acronym seems insufficient. More accurately we could regard lasers as light sources that produce intense coherent light, generated by stimulated emission, and having a positive feedback mechanism [89]. Coherent light emission is the most widely known property of a laser, but since measuring first- and second order coherence of light to determine the laser threshold is impractical, alternative methods are necessary to decide whenever lasing is achieved. Generally, the following characteristics are associated with the presence of laser emission [84, 90, 91]:

- a clear indication of a threshold in the output energy as a function of the input (or pump) energy, with a high lasing efficiency above threshold
- a strong output beam polarization
- spatial coherence of the output beam
- significant spectral line narrowing (coherence in time)
- the existence of laser cavity resonances, or modes

As the acronym LASER suggests, the basic element for laser action is stimulated emission, a process first proposed by Einstein based on thermodynamic considerations. In this process the excited state is stimulated by the interaction of another photon of the same energy to relax radiatively. The crucial point about stimulated emission is the fact that the stimulated photons have the same wavelength, direction

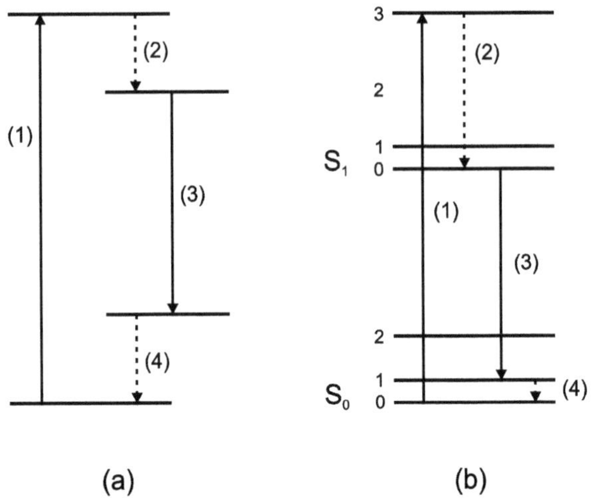

Fig. 1.7 Schematic representation of the energy levels diagrams of (**a**) a generic four-level laser material, and (**b**) the lowest two singlet states in an organic semiconductor. The transitions (*1*) and (*3*) indicate, respectively, optical absorption and emission, whereas transitions (*2*) and (*4*) are thermal relaxations

and phase as the incident ones, leading to the distinctive coherence properties of laser light. In order to achieve a significant contribution from stimulated emission, the Boltzmann equilibrium has to be disturbed in such a way that the population of the upper laser transition state exceeds the one of the lower state. This situation is known as population inversion and can be obtained in a system with more than two energy levels. In a three-level system for example, the upper laser transition state is pumped indirectly via internal conversion from a higher energy state. If the transition rate from this higher energy state to the upper laser transition state exceeds the one between the upper laser transition state and the ground state, the population can be inverted and exploited to give optical amplification by stimulated emission. A more efficient system, however, is a four-level laser system (Fig. 1.7(a)), which allows a further depopulation of the lower laser state by a fast relaxation to the ground state. This results in a higher population inversion and therefore in a more efficient stimulated amplification. The most well-known example of such a system is the He-Ne laser, but also the energy scheme of many luminescent organic materials corresponds to a four-level laser system [92]. The energy levels of a typical organic semiconductor are shown in Fig. 1.7(b). The figure shows the first excited singlet state S_1 and the ground state S_0, both split into a multitude of vibronic states. Photo-pumping from the ground state into the higher located sublevels of the excited singlet state S_1 is succeeded by a very fast non-radiative relaxation to the lowest S_1-level via vibronic relaxation. Subsequently, stimulated emission is possible from this lowest S_1-level to the vibrational states of the ground state, which are then quickly depleted via ground state relaxation.

Generally, a laser device consists of two main parts: an optical amplifier or gain medium and an optical resonator that provides the necessary optical feedback. Both parts can be recognized in Fig. 1.8, which gives a schematic representation of the most common configuration of a Fabry-Perot laser. The gain medium serves as the active material and has the properties to allow light amplification by stimulated emission. It can be of any state: gas, liquid, solid or plasma and is energetically

Fig. 1.8 Schematic representation of laser device structure

pumped by either optical excitation (optically pumped laser) using another laser or a flash lamp or by electrical excitation (electrical pumping). Upon excitation light is generated and photons passing through the gain medium stimulate the emission of more photons, thereby gaining in intensity when population inversion is achieved. The resonator structure then reflects the light backward and forward through the gain medium, realizing an optical oscillation. If the gain (amplification) in the medium is stronger than the resonator losses, stimulated emission can be maintained and a very intense light field is built up. The minimum pump power needed to begin laser action is called the lasing threshold. By virtue of the feedback structure building up laser oscillation is restricted to only resonant modes. Indeed, only photons aligned with the cavity manage to pass more than once through the gain medium and obtain significant amplification. The optical resonator in its most simple form consists of two end mirrors (Fig. 1.8), of which one is only partially reflective, allowing outcoupling of the output laser beam; however there are many alternative feedback designs which can be employed, such as distributed feedback (DFB) structures, micro-cavities, etc. (Sect. 1.3.3).

1.3.2 Motivation for Plastic Lasers

In the last decade, remarkable progress has been made in synthesizing organic semiconductors and in understanding their properties. Currently, organic materials have been incorporated in a number of devices, such as organic light-emitting diodes (OLEDs) [23, 27], organic photovoltaic cells [82] and organic field-effect transistors [93–95]. Notably absent from this list of electronic and optical devices, is the organic diode laser. Organic semiconductor lasers, however, possess numerous advantages over III–V semiconductor lasers, the most important being the higher integration potential on arbitrary substrates, the lower cost and the wider spectral range of possible lasing emission.

There are many photo-physical aspects that make organic semiconductors interesting materials as active laser media. The first is related to the fact that the emission spectra of organic semiconductors is broad, providing possibilities for the fabrication of tunable lasers [96–104] and broad-band optical amplification [105, 106]. In addition, the chemical structure of the organic semiconductor can be changed to give light emission across the visible spectrum. This is illustrated in Fig. 1.9, which shows the emission spectra of three common organic semiconducting laser materials. Moreover, organic materials typically have strong absorption coefficients

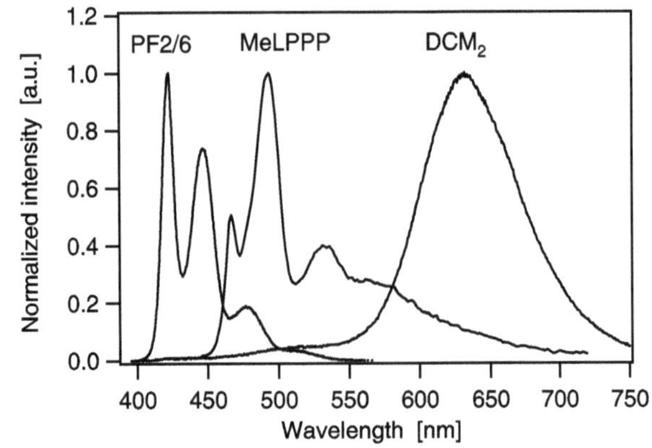

Fig. 1.9 Photoluminescence spectra of three organic semiconducting materials commonly used as laser material

($\sim 10^5$ cm^{-1}) [107], indicating that light can be absorbed over very short distances. Since stimulated emission is closely related to absorption, this means that also strong amplification is possible in organic semiconductors.

Another important property of suitable laser materials is that they have an electronic structure which corresponds to a four-level system, so that the stimulated emission spectrum does not overlap with the ground state absorption spectrum [88]. As mentioned before, most organic semiconductors naturally form a four-level system because structural and vibronic relaxation in the excited state shifts the energy levels [92]. This explains why absorption and emission spectra of organic materials are separated. Due to molecular relaxation of the excited molecules the emission spectrum is red-shifted with respect to the absorption spectrum. In organic semiconductor films, more particularly in conjugated polymer films, there is an additional factor that contributes to the separation of absorption and emission spectra. In conjugated polymers there is typically a high degree of conformational disorder, introducing a distribution of energy levels [108]. Consequently, molecules with a wide range of energies are excited, however, this energy is rapidly transferred to the lowest energy molecules, from which emission occurs [109]. This effect increases further the separation between absorption and emission and is helpful for lasing because it reduces re-absorption at the lasing wavelength.

The fact that high photoluminescence efficiencies can be obtained even in undiluted films [110–113] is another reason why organic semiconductors are attractive laser materials. Considerable effort, however, is needed to control intermolecular interactions in the solid state. The approaches to avoid quenching generally involve increasing the spacing of the chromophores (light-emitting units). For small molecules, this can be done by incorporating the light-emitting dye in a host matrix [41, 114]. In conjugated polymers, bulky side groups are often used to keep the polymer chains apart [40].

1.3.3 *Lasing in Organic Semiconductors*

Historical Overview of Lasing in Organic Materials

The development of organic transistors [115] and organic light-emitting diodes [23, 27] came many years after their inorganic counterparts. In contrast, organic materials have played a significant role in the development of lasers from the beginning. Lasers using organic materials were proposed in the early 1960s [116] and were demonstrated a few years later by Sorokin *et al.* using liquid solutions of organic dye molecules [117]. Due to their broad gain spectrum and wide tuning range [118] commercial dye lasers have been existing for many years and are used for various applications. However, the complex and bulky laser design, requiring regular maintenance, as well as the need to employ large volumes of organic solvents are inherent drawbacks of this technology.

The first solid-state lasers employing organic materials were demonstrated in 1967 by Soffer and McFarland using dye-doped polymers [119] and were followed by the realization of lasing in doped single crystals in 1972 [120] and in pure anthracene crystals in 1974 [121]. Since then, lasing has been demonstrated in a variety of dye-doped solid-state materials [122–125]. In all of these lasers the dye molecules were dispersed in an electrically insulating and optically transparent host matrix, taking prospects for electrical injection away due to the absence of efficient charge transport. Conjugated polymers and organic small molecules, however, may circumvent this problem. Lasing from conjugated polymers was demonstrated for the first time by Moses in 1992 employing a liquid-dye laser configuration [126]. He obtained stimulated emission from the conjugated polymer MEH-PPV[3], which was used in solution. Following this observation, stimulated emission in diluted PPVs was studied extensively and mechanisms that may obstruct lasing in neat films were discussed [127–129]. The first lasing from solid-state conjugated polymers was reported by Hide *et al.* in 1996 [130]. Similar to the solution case, the light-emitting molecules were dispersed in a polystyrene host matrix to prevent the interaction between neighboring polymer chains. Moreover, titanium dioxide (TiO_2) nanoparticles were blended into the MEH-PPV/polystyrene film to enhance scattering. Later that year, four research groups independently observed stimulated emission from neat conjugated polymer films. Graupner *et al.* observed stimulated emission from a poly(para-phenylene)-type ladder polymer using pump-probe experiments [131]. Tessler *et al.* obtained ASE by placing PPV between a dielectric and a semi-transparent silver mirror to form a microcavity [132]. Similar results were obtained by Hide *et al.* [133] and Frolov *et al.* [134] who observed line narrowing in optically pumped waveguides of PPV derivatives. These observations showed for the first time that neat films of conjugated polymers, which are capable of conducting current, could amplify light and that it was not unreasonable to attempt making polymer diode lasers. Apart from conjugated polymers, organic

[3]Poly(2-methoxy-5-(2'-ethyl-hexyloxy)-1,4-phenylene vinylene).

small molecules are also promising laser materials. Optically pumped lasers based on thin films of evaporated organic small molecules were demonstrated by Kozlov *et al.* [90] and Berggren *et al.* [135] in 1997. Since then, lasing action has been studied in different organic semiconductors [136–142] and in a variety of optically pumped structures [143–148], demonstrating the feasibility of organic thin-film materials as active laser media.

Stimulated Emission and Gain in Organic Semiconductors

A prerequisite for lasing is the presence of stimulated emission, which is quantified by the wavelength dependent cross-section for stimulated emission $\sigma_{SE}(\lambda)$. The wavelength dependence of $\sigma_{SE}(\lambda)$ typically resembles the photoluminescence spectrum of the material and is given by:

$$\sigma_{SE}(\lambda) = \frac{\lambda^4 f(\lambda)}{8\pi n^2 c \tau_r}, \tag{1.3}$$

where $f(\lambda)$ is the normalized spectral distribution of the photoluminescence, n is the refractive index of the material, c is the speed of light in free space and τ_r is the radiative lifetime of the involved optical transition. Using this relation, the cross-section for stimulated emission in organic semiconductors can be estimated. For conjugated polymers values in the range of 10^{-16} cm^2 are calculated [149]. This was experimentally confirmed by Haugeneder *et al.* [150] and Holzer *et al.* [151] who measured respectively the cross-section for stimulated emission of LPPP[4] and PmPV-co-DOctOPV[5]. The high cross-sections for stimulated emission measured in conjugated polymers is mainly due to the fact that the π–π^* transitions are fully allowed.

When population inversion is achieved and light travels through an amplifying medium its intensity I increases exponentially with the distance l traveled through the gain medium:

$$I = I_0 \exp[(g - \alpha)l], \tag{1.4}$$

where I_0 is the initial intensity, g is the gain coefficient of the material and α is the loss coefficient. Equation 1.4 clearly shows that lasing is the result of a balance between optical gain and losses in the sample. The gain coefficient is related to the density of excited states N_{exc} and $\sigma_{SE}(\lambda)$ via:

$$g(\lambda) = \sigma_{SE}(\lambda) N_{exc} \Gamma, \tag{1.5}$$

where Γ is the optical confinement factor (i.e. the fraction of the optical wave located in the active layer).

Three main factors determine the overall gain spectrum of an optically pumped organic semiconductor: the gain spectrum of the material, the ground-state absorption and the excited-state absorption. Excited-state absorption is the absorption from

[4] Ladder-type poly(para-phenylene).

[5] Poly(*m*-phenylenevinylene-*co*-2,5-dioctoxy-p-phenylenevinylene).

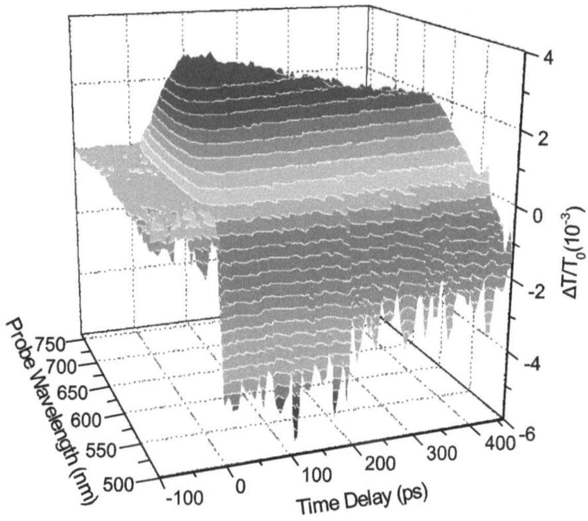

Fig. 1.10 Transient absorption measurement on a 150-nm thick layer of Alq_3 doped with 2%wt DCM_2 deposited on a quartz substrate (Measurements performed by RWTH Aachen)

the radiative S_1-state to higher located energy levels and tends to be problematic for many laser materials. Only when there is no spectral overlap between the gain spectrum of the material and the excited-state absorption, stimulated emission can be obtained. Another competing absorption process is triplet absorption, which arises from those molecules that undergo intersystem crossing and populate the triplet state. The absorption from the T_1-state to the higher located triplet states is particularly important for continuous wave operation, since in this case the population in the T_1-level may accumulate due to the much longer lifetime of the triplet state.

There have been two main approaches to study gain in potential organic semiconductor laser materials. The first is by transient absorption spectroscopy and the second is by measuring amplified spontaneous emission (ASE). Transient absorption measurements or time-resolved pump-probe measurements are a technique that allows ultra-fast measurements of photo-excitations. The sample is typically excited (pumped) by a fast femtosecond laser pulse, which generates photo-excitations, while weaker pulses of light at varying time delays and with varying wavelengths are used to probe the sample. The newly formed photo-excitations then absorb in some parts of the spectrum, while in other parts of the spectrum there is a reduction of the absorption due to the reduced population in the ground state. Therefore, at wavelengths where there is gain, the probe beam will be amplified, whereas at wavelengths where competing absorption processes are dominant, the probe beam is attenuated. An example of such a measurement is shown in Fig. 1.10. The positive transmission change between 550 nm and 730 nm is due to stimulated emission.

The first observation of stimulated emission was reported by Yan *et al.* who performed transient absorption measurements on PPV [152]. In this study, however, the gain was extremely short-lived, which was attributed to the fast formation of intermolecular photo-excitations with strong photo-induced absorption, reducing the net gain in the sample. Since then, several reports have confirmed that high gain is

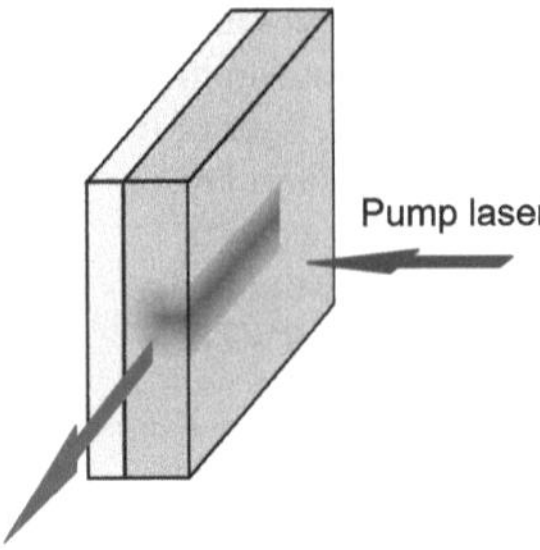

Fig. 1.11 Schematic representation of the typical experimental configuration used for ASE measurements

possible in organic semiconductors [153–156]. MeLPPP[6], for example, has a gain coefficient of around 2000 cm^{-1} [157]. These reports, however, also show that the gain lifetime is usually short, in the order of picoseconds [158]. This represents a challenge for lasers, because a short excited-state lifetime means that a high pumping rate is needed to maintain population inversion [86]. In literature different approaches are suggested to enhance stimulated emission. Denton *et al.* indicated that photo-oxidation must be avoided to obtain maximum gain [129], but also dilution of a polymer in a host matrix can be used to increase the overall gain [127]. For future electrically pumped devices, however, this latter approach might not be useful since charge transport is strongly reduced in diluted systems [159]. In addition, one should keep in mind that the photophysical properties of organic semiconductor films are quite sensitive to packing morphology, which can vary dramatically upon preparation conditions [160, 161].

Although transient absorption spectroscopy provides a powerful means for studying stimulated emission and its competing processes, it is not the easiest method to characterize an organic semiconductor as a laser material. A much easier technique to study gain in organic materials is by measuring amplified spontaneous emission (ASE). This technique involves photo-pumping a slab waveguide in a stripe near the edge of the sample and collecting the edge emission. Typically, the thin-film waveguide is made by depositing the organic semiconductor on a substrate with low refractive index. A schematic representation of the experimental configuration is shown in Fig. 1.11.

When the pump intensity is high enough for the gain to exceed the scattering losses, spontaneously emitted photons are exponentially amplified by stimulated emission according to Eq. 1.4. Since photons whose energy coincides with the spectral position of maximum gain will be amplified more than others a collapse of the emission spectrum, called gain narrowing, is observed. The change in spectral shape upon excitation density is illustrated in Fig. 1.12. This figure depicts the edge-emitted spectra of a 150-nm thick MeLPPP film on glass for two different pump powers. At a laser fluence below the ASE threshold (full line, normalized) a broad emission spectrum spanning from 450 to 700 nm is observed. This is attributed to spontaneous emission of MeLPPP molecules, which are populated in an excited state via optical pumping. Increasing the excitation fluence to higher values

[6]Methyl-substituted ladder-type poly(para-phenylene).

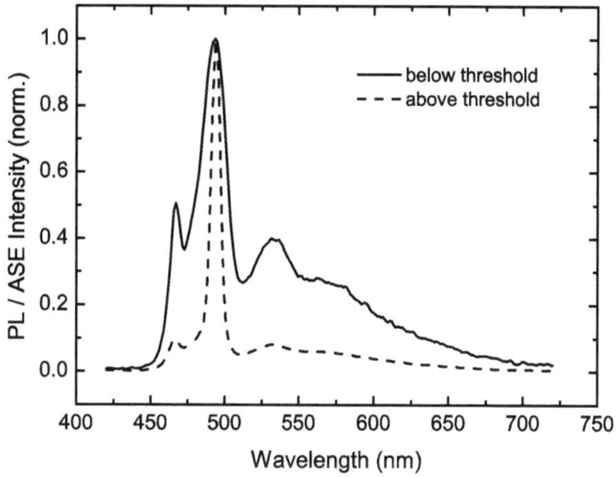

Fig. 1.12 Edge-detected emission spectra of a 150-nm thick MeLPPP film measured at different pump powers. The normalized spectra show a significant spectral gain narrowing at high excitation densities, indicating that ASE has become the dominant deactivation pathway

(dashed line, normalized) results in a spectral narrowing around the peak emission wavelength of 495 nm, which is characteristic for ASE. Here, spontaneously emitted photons, propagating across the optically pumped region, stimulate the emission of excited molecules. In literature ASE has been demonstrated using various organic semiconductors [137, 138, 162–164]. The lowest ASE threshold reported to date is obtained in films based on spiro-SBCz[7] [56].

To extract the gain coefficient from ASE measurements, the variable stripe length method can be applied [165, 166]. In this method the line-narrowed emission is measured as a function of the pump stripe length. The wavelength dependent output intensity, $I(\lambda)$, of the ASE is then given by the expression:

$$I(\lambda) = \frac{A(\lambda) I_p}{g(\lambda)} (\exp(g(\lambda) l) - 1), \tag{1.6}$$

where $A(\lambda)$ is a constant related to the cross-section for spontaneous emission, I_p is the pumping intensity, $g(\lambda)$ is the net gain coefficient, and l is the length of the pumped stripe. By fitting the experimentally measured data to Eq. 1.6 the net gain of the waveguide can be calculated. This method was developed twenty years ago for the inorganic semiconductor cadmium sulfide [165], but has since then been widely applied to determine the gain of organic semiconductors [163, 164, 167, 168]. In addition, this technique provides an easy way to distinguish between ASE and other mechanisms that were proposed to explain the observed spectral narrowing, such as superfluorescence [134, 169] or bi-excitonic emission [170]. If ASE occurs, the spectra should be broad at short stripe lengths and spectral narrowing should occur when the excitation length increases. In contrast, if superfluorescence or bi-excitonic emission is the mechanism of spectral narrowing, the emission spectrum should not depend on the size of the excitation region [85].

The net gain obtained from variable stripe length experiments is determined by waveguide losses, which can arise form self-absorption or scattering. Experiments

[7] 2,7-bis[4-(N-carbazole)phenylvinyl]-9,9'-spirobifluorene.

similar to the net gain measurements can be performed to measure these losses. The pump stripe, having a fixed length, is then progressively moved away from the edge of the sample. Since the emission from the end of the pumped region, I_0, is constant, the emission from the edge of the sample decreases according to Eq. 1.4. Waveguide losses in conjugated polymers typically lie in the range of 3–50 cm^{-1} [86], although even lower losses (<1 cm^{-1}) have been reported in blended organic thin films using cascade energy transfer [135].

Feedback Structures for Organic Semiconductor Lasers

Although ASE has many properties of lasing, such as a distinct threshold and the emission of concentrated, polarized and almost monochromatic light, the above described structures are no lasers because of the absence of resonant modes. Consequently, the output beam obtained in ASE experiments is incoherent. In order to obtain coherent light emission, the gain medium should be incorporated into a resonator, which provides optical feedback. The most common approach to make a resonator is by placing the gain medium between two mirrors. Light is then reflected back and forth between these two mirrors, realizing an optical oscillation. Since positive feedback can only be obtained for those modes that have an integer number of wavelengths in one round-trip of the cavity, each resonator will have a discrete set of resonant optical modes. Lasing occurs when amplification of photons in one of the optical modes is sufficient to overcome the existing losses.

Figure 1.13 features the different geometries that have been explored in the past years to obtain optical feedback. The first solid-state organic semiconductor laser was based on a vertical microcavity formed by a highly reflective dielectric mirror and a semi-transparent silver layer [132] (Fig. 1.13(a)). Following this study, there have been a number of reports about microcavity organic semiconductor lasers based on a broad range of materials [96, 171–173]. The fact that their fabrication is similar to the one of OLEDs and the fact that they emit perpendicular to the substrate, make this type of lasers particularly attractive. The distance traveled by the light during each pass through the gain medium, however, is rather small (mostly <100 nm), resulting in relatively high threshold values [174]. Only if high reflective dielectric mirrors are used at both sides of the microcavity, low laser thresholds can be obtained [172, 173, 175, 176].

Another way to achieve low laser thresholds is by using planar laser resonators where light is waveguided in the organic film via internal total reflection. In such a structure light is waveguided over much longer distances through the gain medium. The most simple way to fabricate a planar waveguide is by cleaving the organic film to obtain reflections at the facets. This is the most common approach used to fabricate low-cost inorganic semiconductor lasers, having very flat facets of typically 30% reflectivity. In contrast to inorganic semiconductors, however, breaking an organic film is rather difficult and does not result in high-reflective, high-quality facets. Nevertheless, lasing in such cavities has been successfully demonstrated [90, 96]. Superior performance can be achieved when feedback is incorporated via a

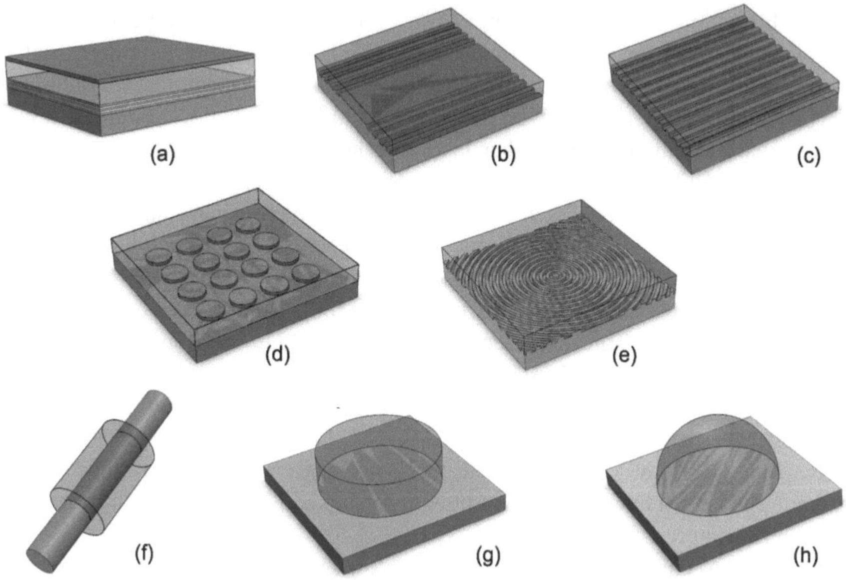

Fig. 1.13 Schematic representation of various resonator structures for optically pumped organic semiconductor lasers: (**a**) vertical microcavity, (**b**) distributed Bragg reflector (DBR), (**c**) distributed feedback (DFB) structure, (**d**) 2D photonic crystal structure, (**e**) circular DFB structure, (**f**) microring, (**g**) microdisk, and (**h**) microsphere

periodic perturbation of the waveguide giving rise to Bragg scattering. If Bragg gratings are placed outside the gain region, replacing the cleaved ends as wavelength selective mirrors, the structure is referred to as a distributed Bragg reflector (DBR) (Fig. 1.13(b)). Berggren *et al.* demonstrated such a laser comprising the small molecule blend Alq_3[8]:DCM_2[9] [114]. The Bragg grating can also be incorporated into the gain region. In this case one has reflection and transmission at every point in the cavity and thus the feedback is distributed throughout the entire length of the cavity. Lasers of this type are known as distributed feedback (DFB) laser, schematically shown in Fig. 1.13(c).

The concept of distributed feedback was introduced in 1971 by Kogelnik *et al.* [177, 178] and relies on Bragg scattering due to a periodic modulation of the refractive index or the gain of the active material. In distributed feedback structures lasing occurs near the Bragg wavelength λ_{Bragg}, which is given by the following relation:

$$\lambda_{Bragg} = \frac{2n_{eff}\Lambda}{m}, \quad m = 1, 2, \ldots. \tag{1.7}$$

Here, n_{eff} is the effective refractive index of the gain material, Λ is the modulation period of the DFB resonator and m is the DFB order number. For $m = 1$, first

[8]Tris-(8-hydroxyquinoline) aluminum.

[9]4-(dicyanomethylene)-2-methyl-6-(julolindin-4-yl-vinyl)-4H-pyran.

order diffraction is obtained, whereas for a second order DFB structure ($m = 2$) the wavelength of the reflected light equals $n_{eff}\Lambda$. In the latter case, laser radiation is also emitted perpendicular to the surface. This can be exploited for obtaining surface-emitting lasers. However, the out-coupled laser radiation represents a loss mechanism, explaining why higher laser thresholds are obtained for second- and higher-order DFB lasers compared to first-order DFB structures [146]. DFB lasers with one-dimensional feedback have been demonstrated using many different organic semiconductors, including derivatives of PPVs [179, 180], polyfluorenes [99, 141, 181] and LPPPs [182, 183], but also small molecule systems have been studied when deposited on top of DFB structures [184–186]. DFB lasers can exhibit very low laser thresholds. Threshold densities as low as 200 nJ/cm^2 have been reported for small molecule blends [184] and 40 nJ/cm^2 for conjugated polymers [148, 187]. DFB lasers also provide possibilities for the fabrication of tunable lasers. By changing either the period of the grating [98, 99, 102, 179, 188] or the thickness of the waveguide [103, 189], one may tune the lasing wavelength over a range of 20–50 nm. The largest tuning range was obtained using an Alq_3:DCM_2 blend, resulting in a tuning range of 115 nm [97].

Apart from one-dimensional DFB structures, also more complicated diffractive resonator structures have been reported. Two-dimensional (2D) feedback can, for example, be obtained in a 2D photonic crystal (Fig. 1.13(d)) with either square [147, 190–193], hexagonal [194] or honeycomb lattices [175]. Concentric circular DFB resonators (Fig. 1.13(e)) [195, 196] can also provide feedback in the plane of the organic semiconductor, although in this case feedback is applied in all directions from a single unique point that is located in the center of the grating. In contrast to their one-dimensional counterparts, 2D lasers exhibit improved output beam quality, lower threshold and higher output efficiencies [181, 190].

Other interesting resonator geometries to fabricate low threshold lasers are microring, microdisk and microsphere cavities (Fig. 1.13(f–h)). These lasers are typically larger in dimension than the vertical microcavities and hence the path traveled by the light through the gain medium in one round-trip is much longer. Consequently, much lower excitation densities are required to achieve sufficient gain [197, 198]. However, due to the long round-trip distance, these lasers have a rather complicated feedback mechanism that is based on a superposition of waveguide and whispering-gallery-mode oscillations [199–204].

Lowering the Lasing Threshold with Energy Transfer

One of the important criteria for active laser materials is the fact that the spectral overlap between stimulated emission and ground state absorption should be minimal in order to reduce self-absorption. Due to the presence of an inherent Stokes shift, the absorption spectrum and the emission spectrum of an organic material are shifted towards each other (Sect. 1.1.3). Most luminescent organic materials, however, still have a slight amount of residual absorption at the peak emission wavelength, reducing the overall gain that can be obtained. This self-absorption problem

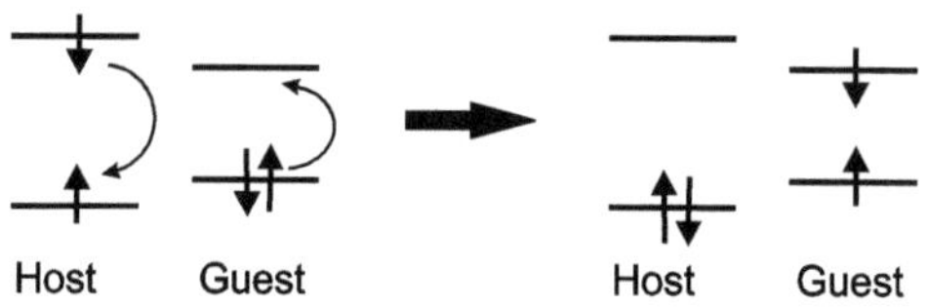

Fig. 1.14 Schematic description of Förster energy transfer

can be alleviated by shifting the emission to longer wavelengths, where residual absorption is smaller. Typically this is done by blending a small amount of a guest molecule or a polymer into a host matrix [90, 114, 135, 143, 205]. In such blends, the excited molecule non-radiatively transfers its energy to a neighboring molecule via long-range dipole-dipole coupling or electron exchange, also known as, respectively, Förster and Dexter energy transfer.

Förster energy transfer describes a radiationless transfer process that is caused by a coulombic interaction between a host and a guest molecule in close proximity (typically <10 nm). This process is shown in Fig. 1.14 and can be formally described by the interaction between the transition dipoles of the two molecules when the intermolecular distance is large enough (>1 nm) [21].

The rate of this energy transfer process k_{ET}^{F} is given by:

$$k_{ET}^{F} = \frac{9000 \ln(10)\kappa^2 \phi_{PL}}{128\pi^5 n^4 N_A \tau_H R^6} \int \lambda^4 f_H(\lambda)\varepsilon_G(\lambda)d\lambda, \tag{1.8}$$

where λ is the wavelength, $f_H(\lambda)$ is the normalized fluorescence spectrum of the host material, $\varepsilon_G(\lambda)$ is the absorption spectrum of the guest in terms of molar absorption coefficient, κ is a polarization factor, ϕ_{PL} is the absolute fluorescence efficiency of the host, τ_H is the radiative lifetime of the host, R is the mean distance between host and guest, N_A is the Avogadro constant and n is the refractive index of the medium. Equation 1.8 clearly indicates that for efficient energy transfer it is important to have a host with a high ϕ_{PL} and a guest with a high absorption coefficient. In addition, the overlap between $f_H(\lambda)$ and $\varepsilon_G(\lambda)$ has to be large. Since the energy transfer rate depends strongly on the distance between the molecules ($\sim R^{-6}$), the integral in Eq. 1.8 is often expressed in terms of the effective Förster radius, R_0:

$$R_0 = \left(\frac{9000 \ln(10)\kappa^2 \phi_{PL}}{128\pi^5 n^4 N_A} \int \lambda^4 f_H(\lambda)\varepsilon_G(\lambda)d\lambda \right)^{1/6} \tag{1.9}$$

which simplifies Eq. 1.8 to:

$$k_{ET}^{F} = \frac{R_0^6}{\tau_H R^6}. \tag{1.10}$$

Hence, the rate of Förster energy transfer between two molecules spaced at a distance $R = R_0$ equals the spontaneous emission rate of the host molecule $1/\tau_H$.

A good example of a Förster energy transfer system is the host-guest system Alq_3:DCM_2, for which an effective Förster radius R_0 of 39 Å has been calculated [96]. The strong overlap between the Alq_3 photoluminescence and the DCM_2 absorption spectra [114], results in efficient energy transfer from Alq_3 to DCM_2.

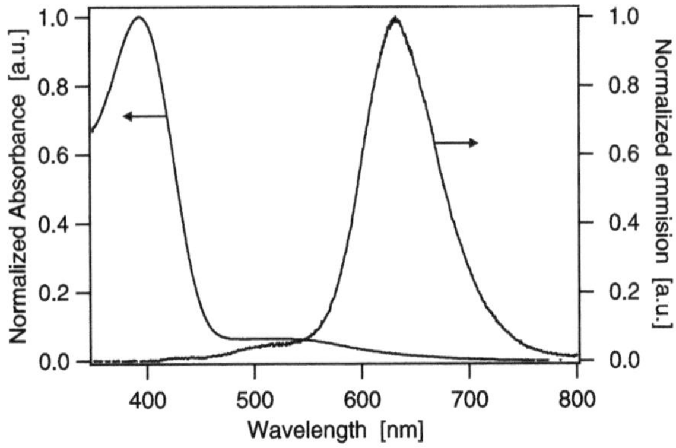

Fig. 1.15 Absorption and emission spectra of a composite Alq_3:DCM_2 thin film, comprising a 2%wt DCM_2 concentration. The absorption and emission spectral band are significantly shifted with respect to each other

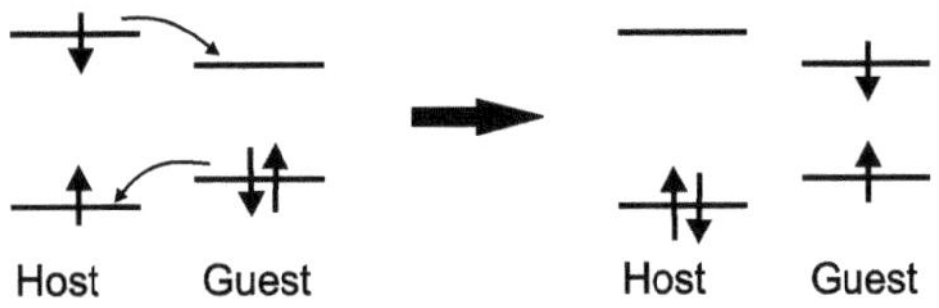

Fig. 1.16 Schematic description of Dexter energy transfer

This is demonstrated in Fig. 1.15. Although excitation was performed at a wavelength of 395 nm, which corresponds to the absorption maximum of Alq_3, the major emission originates from DCM_2, resulting in a very large Stokes shift (240 nm) between the absorption and emission spectral band of the Alq_3:DCM_2 film. Consequently, self-absorption is strongly reduced, making this material system very promising for laser applications. In literature there have been several reports on lasing action based on Förster energy transfer, demonstrating very low lasing thresholds [96, 114, 143, 184].

The Förster energy transfer process described above relies on strong coulombic interactions between the host and the guest. An alternative process for energy transfer, which is based on an electron exchange mechanism, may also be effective in the host-guest system. This process is referred to as Dexter energy transfer and is shown schematically in Fig. 1.16.

Similar to Förster energy transfer, the rate of Dexter energy transfer is proportional to the degree of overlap between the emission spectrum of the host and the absorption spectrum of the guest. However, the Dexter energy transfer rate is calculated using the normalized emission and absorption spectra, removing any dependence of the rate on the guest absorption coefficient. The rate of Dexter energy transfer k_{ET}^D is given by:

$$k_{ET}^D = \frac{2\pi}{h} \exp\left(\frac{-2R}{L}\right) K \int \lambda^4 f_H(\lambda)\varepsilon_G(\lambda)d\lambda, \tag{1.11}$$

where h is Planck's constant, L is the sum of the van der Waals radii of the host and the guest molecules and K is a constant proportional to the orbital overlap between host and guest. Since k_{ET}^D is proportional to $\exp(-2R)$, Dexter energy transfer is very fast for short distances between host and guest molecules (<1 nm), at larger

Table 1.1 Allowed energy transfer mechanism between an excited host molecule and a guest molecule in the ground state. The host molecule is the donor molecule D, whereas the guest molecule is the acceptor molecule A

Förster energy transfer	Dexter energy transfer
${}^1D^* + {}^1A \rightarrow {}^1D + {}^1A^*$	${}^1D^* + {}^1A \rightarrow {}^1D + {}^1A^*$
${}^3D^* + {}^1A \rightarrow {}^1D + {}^1A^*$	${}^3D^* + {}^1A \rightarrow {}^1D + {}^3A^*$

distances, however, its importance decreases rapidly. This is in contrast to Förster energy transfer, which is important at larger distances. Another important difference between Förster and Dexter energy transfer is related to the transitions that are allowed by both processes [21, 206]. While Förster energy transfer requires that the spin of the guest molecules stays unaltered, Dexter energy transfer, does not have a singlet requirement and can efficiently transfer energy between triplet states or between singlet states as long as the total spin symmetry is conserved. Therefore, Dexter energy transfer might still be important at larger distances for those transitions which are forbidden according to the Förster energy transfer process. An overview of mechanisms that are allowed to transfer energy from the excited host molecule to the singlet state of the guest molecule is given in Table 1.1. Both, Förster and Dexter energy transfer can be the mechanism for singlet-singlet energy transfer. However, Förster energy transfer will be the dominant mechanism at low guest concentrations because it is faster over long distances.

1.3.4 Prospects for Electrically Pumped Organic Lasers

Looking at the impressive way in which organic semiconductor materials have become established as new laser materials during the last decade, it is not surprising that main efforts now focus on the development of electrically pumped organic lasers [91]. Electrical pumping is the most convenient way to achieve lasing and is the approach used in inorganic semiconductor diode lasers.

It is expected that high current densities are required to achieve sufficient gain for stimulated emission in organic semiconductors. From optically pumped laser experiments, a lower limit of the current density required to reach lasing threshold can be estimated. Kozlov *et al.* calculated that a current density threshold of 80 A/cm^2 would be needed to obtain stimulated emission in a DFB laser with a DCM[10]-doped Alq_3 emissive layer sandwiched between two cladding layers of Alq_3 [184]. A similar estimation has been made for polymer DFB structures [85]. Several research groups demonstrated that this level of injection can be achieved in organic semiconductors under pulsed operation [207–209]. Recently, current densities as high as 12000 A/cm^2 [210] and 128000 A/cm^2 [211] have been reported in thin films

[10] 4-(dicyanomethylene)-2-methyl-6-[(4-dimethylaninostyryl)-4H-pyran.

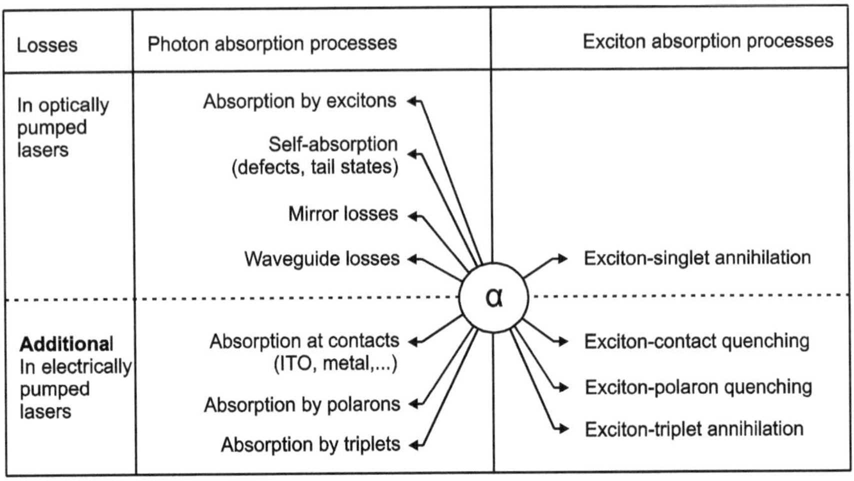

Fig. 1.17 Overview of the possible losses encountered in optically pumped lasers and the additional losses that can be expected in electrically pumped lasers

of copper phthalocyanine. Although this material is not suitable for lasing, it shows that organic semiconductors can sustain extremely high current densities, exceeding the estimated current density threshold by several orders of magnitude. On the other hand, it also demonstrates that the difficulties to realize electrically pumped organic lasers today must be attributed to the presence of additional loss mechanisms compared to optically pumped lasers. Figure 1.17 gives an overview of these additional losses. The figure summarizes the possible losses encountered in optically pumped lasers as well as the additional losses that can be expected in electrically pumped laser devices. A distinction is made between losses of excitons (right column), that make it more difficult to achieve a high N_{exc} in the gain formula (Eq. 1.5), and losses of photons, that result in a higher gain coefficient to be required to achieve stimulated emission: indeed, $g_{min} = \sum \alpha_{photon}$.

Only fluorescent systems, based on singlet emission, can efficiently be pumped by optical excitation. Consequently, the 'exciton-singlet' annihilation process in the right column of Fig. 1.17 refers to singlet-singlet annihilation. At typical threshold concentrations of singlets for lasing, $N_{exc} \sim 10^{17} - 10^{18}$ cm^{-3} [154, 184], this mechanism is not dominant in amorphous organic thin films [85, 212]. Absorption of photons by excitons and self-absorption can be minimized by the choice of material [85]. An appropriate choice of the resonator structure may then further reduce the optical losses. As mentioned in Sect. 1.3.3 mirror losses can be minimized by using diffractive resonator structures. As a result of such optimizations, low losses are possible and optically pumped organic lasers having low laser thresholds can be obtained [146, 184, 187]. On the other hand, to obtain lasing by electrical pumping several other loss mechanisms should still be suppressed.

A first challenge to achieve electrically pumped lasing is to overcome the additional losses due to absorption of photons at the electrical contacts. This is particularly important for organic semiconductor lasers where the resonator is positioned in the plane of the film. In these lasers a long interaction length is present between the light and the gain medium. However, this also means that there is a long interaction with the electrical contacts, which absorb light. Therefore, optical losses in thin waveguide structures including electrodes are very substantial [96]. This was demonstrated by Andrew *et al.*, who reported a considerable increase of the lasing threshold of a MEH-PPV-based organic DFB laser upon inserting a thin silver layer [213]. These detrimental waveguide losses, however, may be reduced by careful optical design [182, 214]. In literature different device architectures have been proposed to reduce the losses caused my metallic layers. A possible strategy is to increase the total thickness of the organic layer stack, by separating the emissive layer from the electrodes by cladding layers that transport the injected electrons and holes towards the active light-emitting layer [84, 215]. In this way the absorption losses at the electrical contacts can be reduced. However, because of the typical low carrier mobility of organic semiconductors [31, 32], the thickness of the organic layers cannot be arbitrary increased. In order to ensure efficient charge transport the total thickness of organic layers is generally limited to less than 200 nm [96]. To circumvent this problem, the use of transparent conductive oxides (TCO) such as indium tin oxide (ITO) [44] and aluminum doped zinc oxide (AZO) [216] has been proposed. A reduction of the organic stack thickness is possible if thin layers of these materials are used as electrical contacts in waveguide structures [96, 217, 218]. This is mainly due to the much lower optical losses in the visible spectral range of TCO layers compared to metallic layers [184]. Görrn *et al.* demonstrated that positioning TCO contacts in the intensity minima of the TE_2 waveguide mode, allows even further minimization of the overlap between the optical mode and the contacts [219]. In addition, Reufer *et al.* reported that it is also possible to use thick, more conductive ITO layers, provided that a cross-linked hole-transport layer is inserted between the highly conductive ITO and the emission layer [220]. Besides conventional LED devices, organic light-emitting transistors (LEOFETs, Sect. 1.2.2) have been suggested as potential laser structures [49, 65, 69, 75, 221, 222]. This configuration makes it possible to position the light-emission zone far away from the metal electrodes and allows minimization of the charge carrier density because of the high field-effect mobility of the carriers. In addition, very high current densities, exceeding several kA/cm^2, have been demonstrated in these devices [63, 69, 75].

Another source of losses associated with electrical pumping is related to the presence of polarons. Electrical excitation involves charge injection. As discussed in Sect. 1.1.1, the injected charges together with their polarization are referred to as polarons. These polarons may cause additional absorption losses [184, 209, 223, 224] and may deplete the exciton population via exciton-polaron quenching [212, 215]. Polaron absorption can be measured by using electro-optical pump and probe experiments and has been observed in polymer [209] as well as in small molecule [184, 207] light-emitting devices. Typically, the polaron absorption spectrum is very broad and covers a wide spectral range. For PPV, for example, the polaron absorption spectrum extends from the ground state absorption band edge to the IR and overlaps with the gain spectrum of the material [225]. Since for low mobility organic semiconductor materials, such as PPV, there are under steady-state conditions much more polarons than singlets [84, 209], polaron absorption will be stronger than stimulated emission, creating a problem for lasing. Several approaches are suggested in literature to avoid the problems related to polaron absorption and polaron quenching [85, 225]. The most obvious solution is to develop a material system where the polaron absorption is shifted with respect to the spectral region of stimulated emission [84, 184]. However, this might not be trivial to find due to the broad nature of polaron absorption. Another possibility is to use organic semiconducting materials with high charge carrier mobility [84, 223]. This would reduce the required charge density in the device, making polaron-exciton quenching and polaron absorption less problematic. A third approach is to operate the device under pulsed excitation [225, 226]. Pulsed operation has several advantages: it avoids heat and stress in the device [227], it allows higher current densities compared to continuous-wave (CW) operation [209], and it makes it possible to separate singlet excitons from polarons in the time domain. This time separation would allow a tremendous reduction of polaron absorption in an electrically driven organic light-emitting device [228].

Besides polarons, current injection also leads to the formation of triplet excitons. As can be seen from Fig. 1.17 these triplet excitations form a third major obstacle to obtain electrically pumped lasing. The inevitable population and accumulation of triplets under electrical excitation results in excessive triplet-state losses preventing lasing [50, 223, 229]. Indeed, if conventional spin statistics apply, the recombination of injected charge carriers leads to the creation of a majority (75%) of non-emissive triplet excitations in the active organic semiconductor layer [36]. Due to their long lifetime, these triplet excitations can act as meta-stable species, which generally have fairly high absorptions to the upper triplet state (triplet-triplet absorption) at the expected fluorescent lasing wavelength [223, 228]. In addition, at high brightness conditions triplets may also severely quench singlet excitons, reducing N_{exc} [50, 212, 215]. Similar to the approach for reducing the amount of polarons in the device, pulsed operation can also be used to separate singlets from triplets in the time domain [228]. By applying pulses which have a pulse duration longer than the singlet lifetime (typically 1 ns) and shorter than the lifetime of triplet excitons (milliseconds to seconds for fluorescent materials) [230], and assuming the pulse repetition rate is low, the accumulation of triplet excitons can be reduced. In this way triplet state losses such as singlet-triplet annihilation and triplet-triplet absorption might be suppressed. The use of appropriate triplet scavengers is also proposed as an alternative way to reduce triplet accumulation [50].

In fact, the critical issues outlined above are mainly related to the low charge carrier mobility of organic semiconductors. The low mobility makes it difficult to obtain high current densities, it prevents the use of thick organic layers to reduce absorption at the metallic contacts and it increases the number of charge carriers needed to generate population inversion in the device. Hence, the use of new materials with higher mobilities would help with each of these issues. Combining a high mobility with simple processing and high photoluminescence efficiency is, however, a challenging project. Therefore, an intermediate solution based on indirect electrical pumping has been proposed, allowing compact and low-cost tunable laser devices [229, 231]. In these devices solid-state inorganic laser diodes are used to pump an organic laser structure optically, circumventing problems associated with electrical pumping. Several demonstrations of indirect electrically pumped lasing in hybrid laser devices have been reported [99, 231, 232].

Recently, spectrally narrowed emission from electrically pumped organic semiconductor films has been demonstrated. Yokoyama *et al.* observed spectrally narrowed emissions from the edges of electrically pumped OLEDs comprising BSB-Cz[11] [233]. These emission spectra corresponded well to the cutoff wavelength of the waveguide structure. In addition, the edge emission intensity of the TE mode showed a superlinear dependence on the current density. The combination of both effects has been attributed to the occurrence of light amplification. Tian *et al.*, however, reported that the apparent "optical gain" associated with spectrally narrowed edge emission from OLEDs, might also result from misalignment of the propagating leaky waveguide modes and the collecting optics [234]. Very recently, Liu *et al.* reported lasing action in an electrically pumped microcavity device [176]. The device showed a threshold current density of 860 mA/cm^2 under room temperature pulsed operation. However, if the observed emission is real laser emission is still under debate. The rather strange spectral behavior of the device as function of the current, the limited spectral narrowing and the fact that spacial coherence can also be the result of other factors, such as a small pumping area or a decrease in the current spreading, might suggest that the observed emission origins from a microcavity-OLED rather than from a laser device.

1.4 Outline

In this first chapter, the basic principles of organic optoelectronics were highlighted and the state-of-the-art light-emitting devices as well as the challenges that have to be overcome to achieve electrically pumped lasing were discussed. The following chapters of this book focus on the design of new device and material concepts for organic light-emitting devices. Main attention is thereby given to the control of the triplet density and to the possibility to achieve high current densities combined with reasonable efficiencies and reduced absorption at metallic contacts. In this respect the proposed concepts offer new prospects for the realization of solid-state high-brightness organic light sources.

Chapter 2 starts with an overview of the organic materials and the experimental techniques that were used during this work. Sample fabrication as well as the different methods used to characterize the devices are discussed.

In Chap. 3, an OLED with field-effect electron transport is proposed as a new device architecture complementary to the list of existing electroluminescent devices. The device is a hybrid structure between a diode and a field-effect transistor. It allows minimization of the optical absorption losses and the charge carrier density in the device, while at the same time high current densities can be achieved. Device

[11] 4,4'-bis[(N-carbazole)styryl]biphenyl.

fabrication and operation are discussed in detail. The electrical characteristics and the opto-electronic performance of the device are measured and are compared with numerical simulations.

Improving the device performance is the subject of Chap. 4. Two diperfluorohexyl-quater thiophene derivatives are investigated for use as the electron-trans porting material in OLEDs with field-effect electron transport. First, the growth conditions of both materials are optimized and high-mobility field-effect transistors are fabricated. Using the optimized growth conditions OLEDs with field-effect-assisted electron transport are fabricated and their electrical characteristics are measured. A comparison between the performance of different devices comprising various electron-transporting materials is given at the end of this chapter.

Chapter 5 presents methods to control the triplet concentration in electrically driven devices. A first method to control triplet accumulation is by operating the device under pulsed excitation. The behavior of OLEDs with field-effect-assisted electron transport is analyzed under these conditions and their performance is compared to conventional OLEDs comprising the same set of materials. Experiments to identify the limiting factors with respect to the pulse width dependence of the light-emitting devices are performed. An alternative way to reduce the triplet population is investigated in the second part of this chapter. Polymer films doped with a nonvertical triplet scavenger are studied by time-resolved photoluminescence measurements. The results are compared to those obtained by using a vertical triplet acceptor, which also possesses suitable energetic levels to scavenge host triplets.

Triplet dynamics and charge carrier trapping in triplet-emitter doped conjugated polymers are the focus of Chap. 6. Phosphorescent properties and charge carrier trapping are studied by means of time-resolved photoluminescence and thermally-stimulated luminescence techniques. In this chapter also the route of using triplet-emitters as light-emitting species in the active layer of an organic laser is investigated. Three different phosphorescent compounds are carefully checked, however, no stimulated emission could be observed.

In Chap. 7, the value of OLEDs with field-effect-assisted electron transport as laser device configuration is discussed. The reduced absorption losses compared to conventional OLED structures are analyzed. Amplified spontaneous emission measurements and optically pumped lasing experiments are performed on organic layer stacks including hole- and electron transporting layers used in OLEDs with field-effect-assisted electron transport. In addition, the different possibilities to incorporate an optical feedback mechanism in the device are discussed.

Finally, conclusions are drawn and some outlook for future research is given in Chap. 8.

Chapter 2
Materials and Experimental Techniques

This chapter discusses the different preparation and characterization techniques that were used during this work. In Sect. 2.1 we focus on the materials and the deposition techniques that were exploited. In addition the fabrication of nanostructured substrates is briefly discussed. The different techniques that were used to characterize the samples are subject of Sect. 2.2.

2.1 Sample Fabrication

2.1.1 Materials Used in this Work

The molecular formulas of the organic materials that were used in this work are represented in Fig. 2.1. The correct chemical designation of each organic compound can be found in Table 2.1. PTCDI-$C_{13}H_{27}$, Alq_3, PαMs, COT and anthracene, were purchased from Sigma-Aldrich, CNPPP and DCM_2 from H.W. Sands Corp., DFH-4T and DFHCO-4T from Polyera Corp. and $Btp_2Ir(acac)$ from American Dye Source Inc. PTAA is received from Merck, whereas F5Ph and GDP16b came from the group of Prof. L. De Cola. MeLPPP and PF2/6 were synthesized by the group of Prof. U. Scherf. Most of these materials were used as received. Only PTCDI-$C_{13}H_{27}$ and Alq_3 were purified once by train sublimation before loading into an ultra-high vacuum system ($p = 10^{-8}$ torr). The lowest unoccupied molecular orbital (LUMO) and the highest occupied molecular orbital (HOMO) of most of these molecules are depicted in Fig. 2.2.

Prior to the deposition of organic materials, the substrates were cleaned with solvents (acetone and isopropanol) and exposed to an UV-ozone ambient for 15 minutes. Processing steps after UV-ozone treatment were usually carried out in a dry N_2 glovebox (<1 ppm O_2, <5 ppm H_2O) or in ultra-high vacuum. Both systems are attached to each other, making transport of samples from high-vacuum to the glovebox and vice versa possible without exposure to ambient atmosphere.

S. Schols, *Device Architecture and Materials for Organic Light-Emitting Devices*,
DOI 10.1007/978-94-007-1608-7_2, © Springer Science+Business Media B.V. 2011

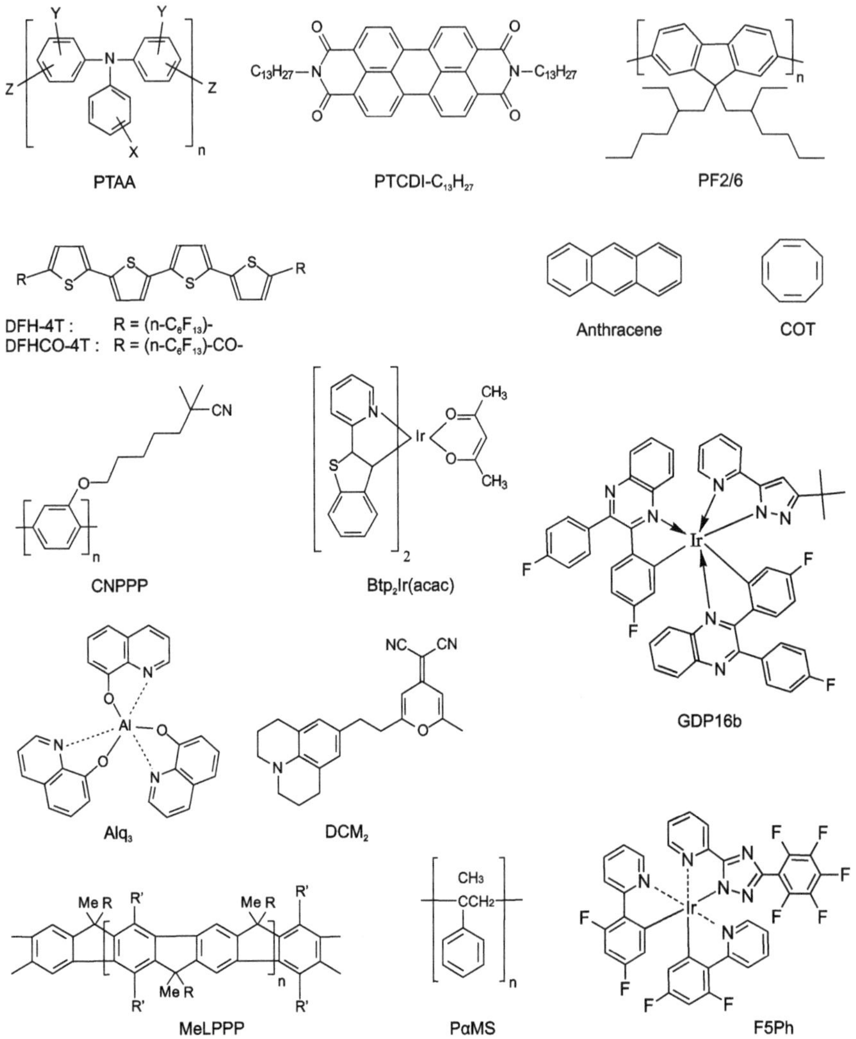

Fig. 2.1 The molecular structure of the organic materials used in this work

2.1.2 Deposition Techniques

Organic semiconductor films can be fabricated using various deposition techniques. If the molecule is soluble in an organic solvent, deposition techniques such as spin-coating, drop-casting, spray-coating and inkjet-printing can be used. Polymer semiconductors are usually deposited is this way. Organic small molecules on the other hand, are generally not very soluble in common solvents and are therefore typically deposited using organic molecular beam deposition (OMBD) or by organic vapor deposition. In these techniques the organic material is heated above the evaporation

Table 2.1 Chemical description of the organic compounds shown in Fig. 2.1

Abbreviation	Chemical description
Alq_3	tris(8-hydroxyquinoline) aluminum
$Btp_2Ir(acac)$	bis(2-(2'benzothienyl)pyridinato-N,$C^{3'}$)(acetylacetonate)iridium(III)
CNPPP	2-[(6-Cyano-6-methylheptyloxy)-1,4-phenylene] copolymer
COT	1,3,5,7-cyclooctatetraene
DCM_2	4-(Dicyanomethylene)-2-methyl-6-(julolindin-4-yl-vinyl)-4H-pyran
DFH-4T	α,ω-diperfluorohexyl-quaterthiophene
DFHCO-4T	5, 5'''-diperfluorohexylcarbonyl-2,2':5',2'':5'',2'''-quaterthiophene
F5Ph	bis((2,4-difluoro)phenylpyridine)-(2-(1,2,4-triazol-3-pentafluorophenyl)-pyridine)iridium(III)
GDP16b	2,3-Bis(4-fluorophenyl)quinoxaline(3-tert-butyl-5-(2-pyridyl)pyrazole)-iridium(III)
MeLPPP	methyl-substituted ladder-type poly(para-phenylene)
PαMs	poly-α-methylstyrene
PF2/6	poly(9,9-di(ethylhexyl)fluorene)
PTAA	Poly(triarylamine)
PTCDI-$C_{13}H_{27}$	N,N'-ditridecylperylene-3,4,9,10-tetracarboxylic diimide

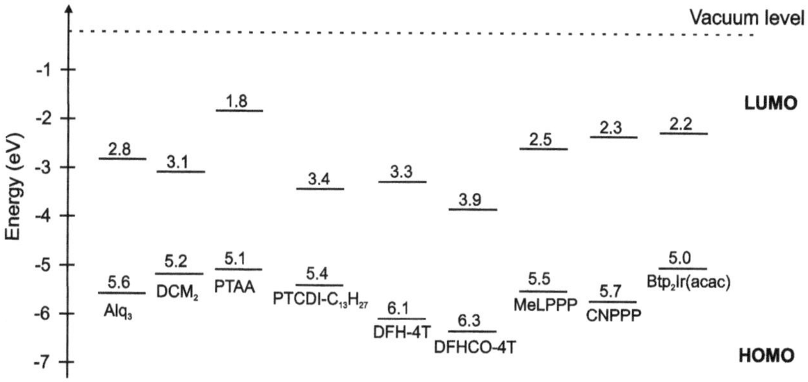

Fig. 2.2 The HOMO and LUMO energy levels of most organic semiconducting materials used in this work [235–240]

temperature and transported towards the substrate in vacuum or with a carrier gas. In this work organic thin-films were prepared using OMBD or spin-coating. Both techniques will be briefly discussed below.

Table 2.2 Typical processing conditions for the organic compounds of Fig. 2.1 that were deposited via spin-coating

abbreviation	concentration mg/ml	spin-speed (rmp)	T_{bake} (°C)	thickness (nm)
CNPPP	10 (chlorobenzene)	1000	–	100
MeLPPP	15 (toluene)	1000	–	150
PαMs	0.86 (toluene)	4000	120 (1 min.)	5
PF2/6	10 (toluene)	1000	–	100
PTAA	45 (toluene)	1000	110 (20 min.)	50

Spin-Coating

The most commonly used deposition technique for soluble organic semiconductors is spin-coating. The substrate, which is centered on the chuck of a spin-coater, is covered with a solution containing the dissolved organic compound. Upon revolving the sample, centrifugal force spins off the fluid and a uniform film is created. Afterwards, the persistent solvent can be removed by baking the sample on a hotplate at elevated temperatures. The resulting film thickness depends on the molecular weight of the organic material, the concentration of the solution and the spin-speed. For the organic compounds of Fig. 2.1 that were deposited by spin-coating, the typically used concentration, solvent, spin-speed, baking temperature T_{bake} and the resulting film thickness are summarized in Table 2.2. The layer thickness was determined using a Veeco Dektak V200-Si stage profiler or by spectroscopic ellipsometry (Sopra GESP-5).

Organic Molecular Beam Deposition

A disadvantage of solution processing is the fact that the fabrication of heterostructures is limited by the demand for orthogonal solvents. This problem can be circumvented by deposition of the organic compounds by OMBD. OMBD is a standard technique to deposit small molecular weight organic materials. The organic material is loaded into a (ultra) high vacuum chamber inside a container (boat or crucible) which can be heated. When the organic material is heated above its evaporation temperature, molecules are sublimed from the container. The molecular beam is directed towards the substrate, where molecules can condense and form a layer. In this way, films of only a few monolayers up to micrometers thickness can be obtained with high accuracy. Depending on the deposition conditions (deposition flux and substrate temperature) the growth of the organic materials can be controlled and an amorphous or polycrystalline film is formed.

At imec, four ultra-high vacuum chambers ($p = 10^{-8}$ torr) are available for OMBD of organic materials and evaporation of metals. The four chambers are connected to a central transfer chamber, which is for his part connected to a N_2-filled glovebox by means of a loadlock. Each of the four evaporation chambers is equipped

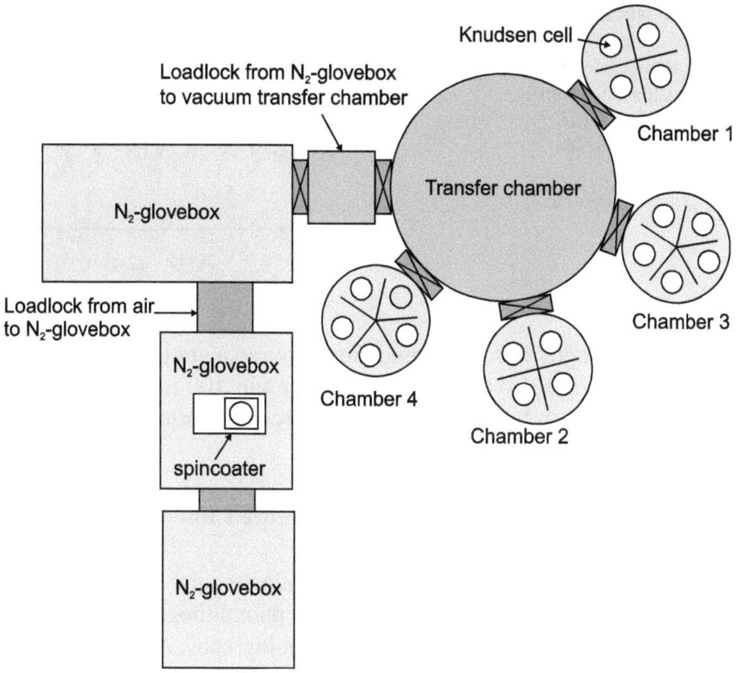

Fig. 2.3 Ultra-high vacuum cluster system with attached N_2 gloveboxes for vacuum deposition of organic materials and metals

with several Knudsen cells, containing a crucible filled with an organic material or a metal. The deposition rate can be controlled by the temperature of the Knudsen cell and is in-situ measured by a calibrated water-cooled quartz crystal monitor. The temperature of the substrate can also be independently controlled. Every cell has a shutter that can be used to block the evaporated material flux. In addition, the flux reaching the substrate can be blocked by another shutter located close to the substrate itself, which is particularly useful to calibrate the deposition rate. PTCDI-$C_{13}H_{27}$, Alq_3, DCM_2, DFH-4T, DFHCO-4T, $Btp_2Ir(acac)$ and GDP16b were typically deposited using OMBD. A schematical illustration of the ultra-high vacuum cluster system available at imec is shown in Fig. 2.3.

2.1.3 Fabrication of Photonic Feedback Structures

As described in Chap. 1 (Sect. 1.3.3) low threshold organic semiconductor lasers can be obtained using diffractive resonator structures. Distributed feedback relies on a periodic modulation of the refractive index or the gain of the active material and can be achieved by corrugating the substrate with a periodic height modulation. In close collaboration with IBM Research Zurich and AMO GmbH in Aachen different photonic feedback structures have been fabricated. For structure definition two main

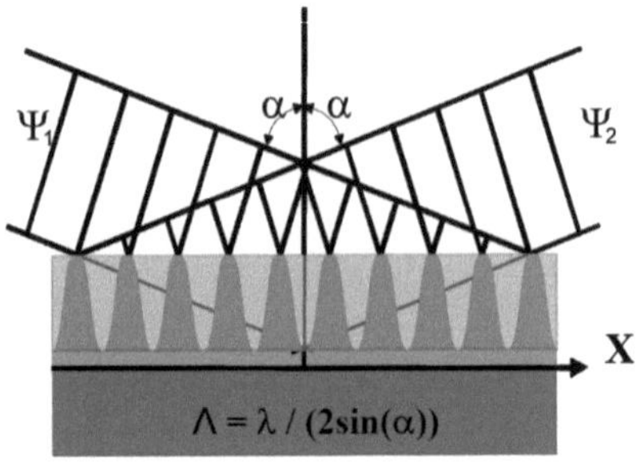

Fig. 2.4 Schematical illustration of 2-beam interference lithography. An interference pattern between two coherent light waves is transferred to a photoresist layer

lithography technologies have been used: first, large area gratings have been realized by interference lithography (IL) and second, for sub 100 nm resolution in dense structures electron beam lithography (EBL) has been performed.

Fabrication of Linear Gratings Using Interference Lithography

Interference lithography is a technique for patterning regular arrays of fine features, without the use of complex optical systems or photolithography masks. The basic principle is the same as in interferometry or holography. An interference pattern between two or more coherent light waves is set up and recorded in a layer (photoresist). This interference pattern consists of a periodic series of fringes representing intensity minima and maxima. Upon post-exposure photolithographic processing, a photoresist pattern corresponding to the periodic intensity pattern emerges. For 2-beam interference, as depicted in Fig. 2.4, the period is given by $\lambda/(2\sin(\alpha))$, where λ is the wavelength and α is the angle between the interfering wave and the substrate normal. The minimum period achievable is half of the wavelength.

Large area linear gratings have been realized with laser interference lithography. Typically, the exposures were carried out on 1.8 μm thick SiO_2 layers thermally grown on Si substrates. On top of these substrates a two layer resist, consisting of a bottom anti reflection coating (BARC) layer to suppress substrate reflectance and a chemical amplified resist, was deposited. Exposure was performed at a wavelength of 266 nm and by exact adjustment of the exposure dose the desired line width of the grating could be achieved. In this way, different gratings with different line width and period could be realized. A scanning electron microscopy image (SEM) of the resist structure of a linear grating (line width 160 nm and period 325 nm) after development is shown in Fig. 2.5.

After lithography the resist pattern was transferred into the SiO_2 by dry etching. An inductively coupled plasma reactive ion etching (ICP-RIE) tool was used to perform this etching. In a first step, the BARC layer of the samples exposed by IL was etched using O_2 plasma. Next, an etching process based on C_4F_8[1] and helium (He) chemistry made it possible to transfer the lithographically defined structures into

[1]Octafluorocyclobutane.

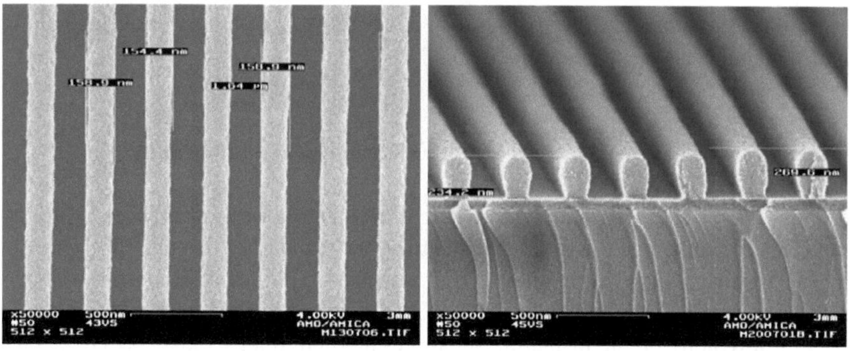

Fig. 2.5 Top view (*left*) and cross-sectional view (*right*) SEM images of 160 nm wide lines realized by IL. (Images received from AMO GmbH)

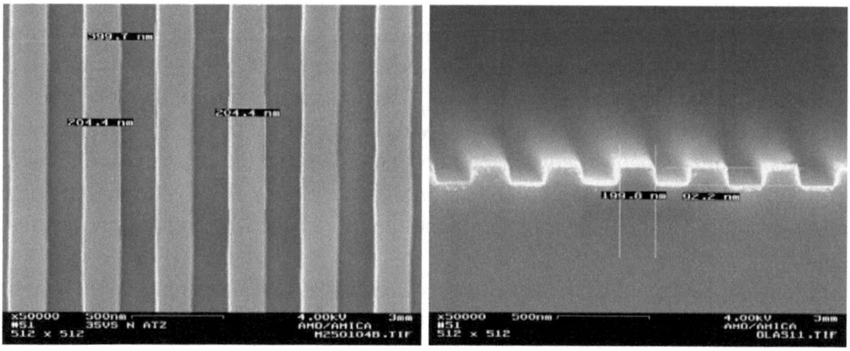

Fig. 2.6 (*Left*) Top view SEM micrograph of a fully fabricated linear grating resonators in SiO_2, and (*right*) cross-sectional SEM image of the same grating. The 90 nm deep trenches exhibit a profile angle of better than 80°. (Images received from AMO GmbH)

the SiO_2 layer. Depending on the actual design of the pattern a precise adjustment of many process parameters like ICP/RF power, C_4F_8/He mixture, etching temperature etc. was necessary to achieve a satisfying etching result. As an example for the dimensional stability of the optimized etching process a SEM micrograph of a completely fabricated grating is shown in Fig. 2.6 (left). The grating had a desired line width of 200 nm. The deviation between the designed pattern size and the realized pattern size is in the range of a few nm. This demonstrates the potential of IL and ICP etching to fabricate defined linear gratings with a line width accuracy of a few nm. Further, IL allows the fast production of large area gratings with a wide range of line widths ranging from 80 nm to 500 nm. This flexibility makes IL an ideal lithography technology for testing different materials and different grating structures. Besides the dimensional stability between the lithographic and the etched structures, the steepness of the etched trenches is also an important parameter, defining the quality of the etching process. To determine the profile of the gratings the cross-section has been investigated using SEM. Figure 2.6 (right) shows the cross-sectional SEM image of the grating displayed in Fig. 2.6 (left). A profile angle of better than 80° could be determined.

Fabrication of Linear Gratings Using Electron Beam Lithography

Although IL is a powerful technique that allows fast fabrication of large area gratings, other technologies with higher contrast are mandatory for sub 100 nm resolution. Electron beam lithography (EBL) was used in this case. Exposure was performed at AMO GmbH with a Leica EBPG5000 system. To ensure accurate electron beam lithography on the insulating substrates a 40 nm thick Ti layer was deposited as intermediate layer between the SiO_2 surface and the electron sensitive resist. The Ti layer prevented charging during the exposure and acted as a hard mask for the following etching processes. To resolve fine structures an EBL process with high contrast value is necessary. The negative tone electron sensitive resist HSQ[2], prebaked at moderate temperature, fulfils this requirement. The resist was spin-coated on top of the substrate resulting in a thickness of about 200 nm, which is sufficient to act as an efficient etching mask. After electron beam exposure the development was carried out in highly concentrated TMAH[3]. As shown in Fig. 2.7 (left), clearly resolved 80 nm trenches between 240 nm wide lines can be obtained in this way. Figure 2.7 (right) shows the cross-section of a developed HSQ pattern. The steep resist structures without any broadening at the foot of the line demonstrate its high quality.

Using the 200-nm thick HSQ pattern as an etch mask, the structures could then be transferred into the SiO_2 substrate using the same ICP-RIE tool as for the samples exposed by IL. First, a BCl_3[4] chemistry based etching process was performed to etch

Fig. 2.7 (*Left*) Top view SEM image of a linear grating in HSQ with line to space ratio of 1:3, and (*right*) Cross sectional SEM image of a HSQ resist pattern. (Images received from AMO GmbH)

---

[2]Hydrogen silsesquioxane.

[3]Tetramethylammonium hydroxide.

[4]Boron trichloride.

the Ti layer. Second, the main etching process based on C_4F_8 and He chemistry was applied.

2.2 Device Characterization

After fabrication the samples and devices were characterized using various techniques. Basic characterization was performed using atomic force microscopy, ellipsometry, transmission spectroscopy and absorption measurements. Some other measurement techniques used to characterize our samples are briefly discussed in this section.

2.2.1 *Transistor Measurement and Parameter Extraction*

The electrical characteristics of organic thin-film transistors were measured in an inert N_2 atmosphere using an Agilent 4156C parameter analyzer. Standard x-y-z control probes (Karl Suss, PH120) were used to connect three source-measure units (SMU) to the device. Two major types of transistor measurements were usually performed. First, the transistor output characteristics were measured, meaning that a fixed gate voltage was applied to the gate electrode, while varying the drain voltage. This results in I_{ds}–V_{ds} measurements for different gate voltages as illustrated in Fig. 2.8(a) for an n-type organic thin-film transistor. Two regimes can be distinguished: the linear regime, where I_{ds} increases with V_{ds}, and the saturation regime, where I_{ds} is almost independent of V_{ds}. Second, the I_{ds}–V_{gs} or transfer characteristics were measured. This type of experiment measures the drain current as a function of the gate voltage. Depending on the value of the drain voltage, the transistor operates either in the linear or in the saturation regime during the measurement. An

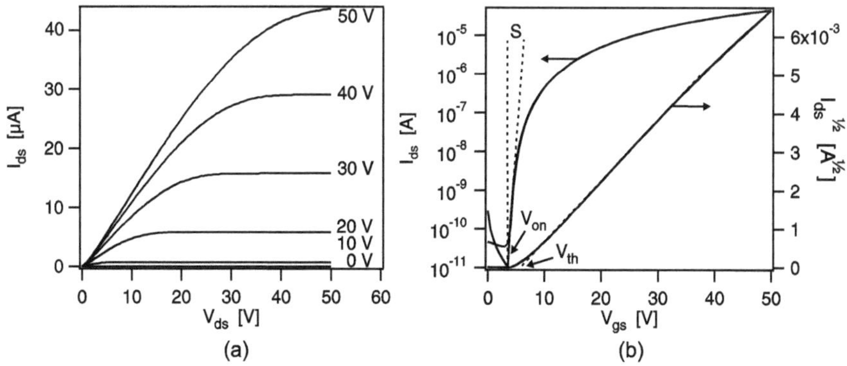

Fig. 2.8 (**a**) Typical output characteristics, and (**b**) transfer characteristics of an n-channel organic thin-film transistor with PTCDI-$C_{13}H_{27}$ as the organic semiconductor ($W/L = 2000/130$)

example of a transistor transfer characteristic, measured in the saturation regime is shown in Fig. 2.8(b).

The transfer characteristics are very useful for extracting the basic transistor parameters. Assuming the gradual channel approximation, the current-voltage characteristics can be analytically described by:

$$I_{ds} = \frac{W}{L}\mu C_{ox}\left(V_{gs} - V_{th} - \frac{V_{ds}}{2}\right)V_{ds} \quad (|V_{ds}| < |V_{gs} - V_{th}|), \tag{2.1}$$

$$I_{ds} = \frac{W}{2L}\mu C_{ox}(V_{gs} - V_{th})^2 \quad (|V_{ds}| > |V_{gs} - V_{th}|), \tag{2.2}$$

where W is the channel width, L is the channel length, μ is the mobility and C_{ox} is the capacitance of the gate dielectric per unit area. From Eq. 2.2 it is obvious that $\sqrt{I_{ds}}$ is linearly proportional to V_{gs} and that the slope of this line is proportional to the mobility μ. The saturation field-effect mobility can thus easily be extracted from this curve. In addition, extrapolating the linear part of $\sqrt{I_{ds}}$ to zero current yields the threshold voltage V_{th}, as shown in Fig. 2.8. Plotting the transfer characteristics on a semi-logarithmic scale allows extraction of two other transistor parameters: V_{on}, defined as the voltage at which the drain current abruptly increases above a defined low off-current level, and the subthreshold slope, S, which depends on the gate dielectric capacitance and the trap states at the interface [77].

The total capacitance of the gate dielectric per area can be calculated based on the relative dielectric constant ε_r and the thickness of the dielectric layer t. Capacitance-voltage measurements pointed out that the relative dielectric constant of sputtered SiO_2 is $\varepsilon_r = 5$, which is somewhat larger than the value obtained for thermally grown SiO_2 ($\varepsilon_r = 3,9$). The measured ε_r for PαMS is 2.5.

2.2.2 Characterization of OLEDs

The fabricated OLEDs were measured in a N_2-filled glovebox using standard probe needles, which were connected to the SMUs of an Agilent 4156C parameter analyzer. Simultaneous to current-voltage measurements the emitted optical power was detected. This was done using a calibrated integrated sphere (SphereOptics Hoffman GmbH). Because of the large sample size and the impossibility to wire-bond our light-emitting devices, the sample was positioned on top of this integrated sphere. This introduced a small error on the measured light intensity. Assuming a Lambertian dependence of the emitted light [241], i.e. $I = I_0 \cdot \cos\theta$, a 5% loss of the total optical power of the light-emitting device was calculated. The measured light intensity P_{meas} is thus only 95% of the actual optical power P_{tot} of the devices:

$$P_{meas} = 0.95 P_{tot}. \tag{2.3}$$

An additional consequence of this measurement setup is that light which is waveguided in the sample substrate is not taken into account. A photograph of the setup is shown in Fig. 2.9 (left).

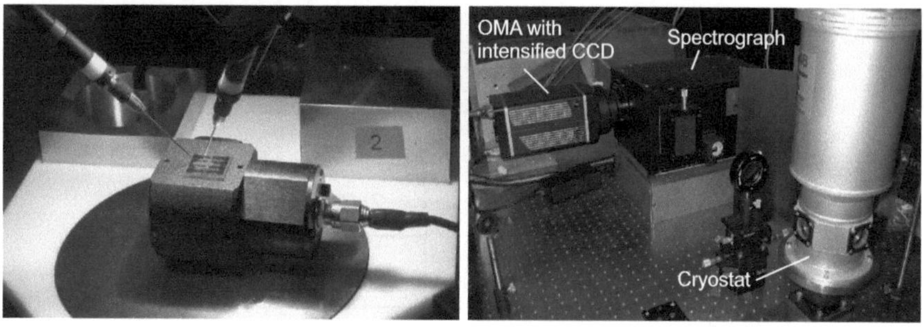

Fig. 2.9 (*Left*) Photograph of the setup used to characterize OLEDs, and (*right*) Photograph of the setup employed to determine the spectral characteristics of OLEDs

To determine the spectral characteristics of our OLEDs, the emitted light was detected by means of an optical multichannel analyzer (OMA) in conjunction with charge coupled device (CCD) (Fig. 2.9 (right)). These measurements were performed at room temperature in a cryostat or by using a small N_2 container to prevent photo-oxidation. The external quantum efficiency η_{ext} could then be calculated based on luminance, electroluminescence spectrum and current. η_{ext} is defined as the number of emitted photons per electron-hole pair injected into the active light-emitting layer, i.e.

$$\eta_{ext} = \frac{P_{tot}/E_{phot}}{I/q}, \tag{2.4}$$

where E_{phot} is the average photon energy, I is the injected current and q is the elementary charge. If E_{phot} is given in eV, the external quantum efficiency reduces to:

$$\eta_{ext} = \frac{P_{tot}[\mathrm{W}]}{I[\mathrm{A}]E_{phot}[\mathrm{eV}]}. \tag{2.5}$$

The average photon energy E_{phot} is calculated by:

$$E_{phot} = \frac{\int (hc/\lambda) f(\lambda) d\lambda}{\int f(\lambda) d\lambda}, \tag{2.6}$$

where λ is the wavelength, h is Planck's constant, c is the speed of light in free space and $f(\lambda)$ is the emission spectral distribution. Since the emission spectrum of OLEDs is not always symmetric around the emission peak, this energy often differs from the energy corresponding to the peak emission wavelength.

2.2.3 Time-Resolved Photoluminescence Measurements

Time-resolved photoluminescence spectroscopy is a characterization technique where the emission of a sample is monitored as a function of time after excita-

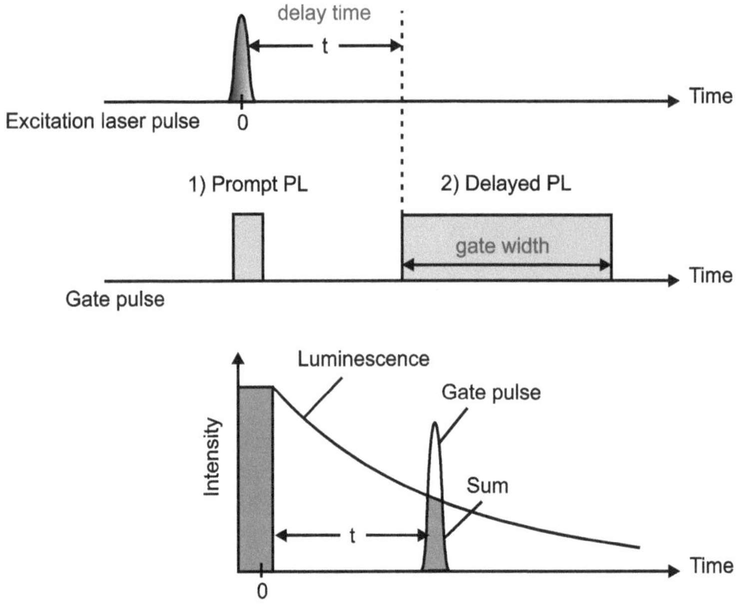

Fig. 2.10 Principle of time-resolved photoluminescence measurements

tion. This makes it possible to separate prompt fluorescence from delayed emission. Delayed emission originating from the radiative decay of singlets is referred to as delayed fluorescence, whereas the emission from triplets is called phosphorescence. Time-resolved spectroscopy also allows determination of the decay kinetics of light-emitting materials. By recording the emission spectra over a wide range of delay times after the laser pulse excitation, the decay of the emission intensity can be measured. The lifetime of the light-emitting species can then be obtained by fitting this curve. The principle of time-resolved photoluminescence spectroscopy is illustrated in Fig. 2.10.

The setup used at imec to perform time-resolved photoluminescence measurements is depicted in Fig. 2.11. A nitrogen laser (Spectra-Physics) with pulse duration of 4 ns operating at 10 Hz was used for optical excitation at 337 nm. The emission spectra were recorded using a triple-grating monochromator (Acton SpectrPro 2300i) coupled to an intensified CCD camera (PI-MAX from Princeton Instruments) with a time gated, intensified diode array detector, which was synchronized by the electrical trigger of the laser. The detection window could be selected between 1 ns and 25 ms. A variable delay of 75 ns to 25 ms after optical excitation allowed the detection of weak delayed luminescence after the intense prompt fluorescence. The spectra were typically accumulated by averaging over 100 to 300 pulses in order to increase the signal-to-noise ratio. To prevent photo-oxidation the sample was loaded into a temperature regulating nitrogen cryostat, which allowed spectroscopic measurements within a temperature range from 77 K to 350 K. Decay kinetics of phosphorescent materials could be studied using this setup.

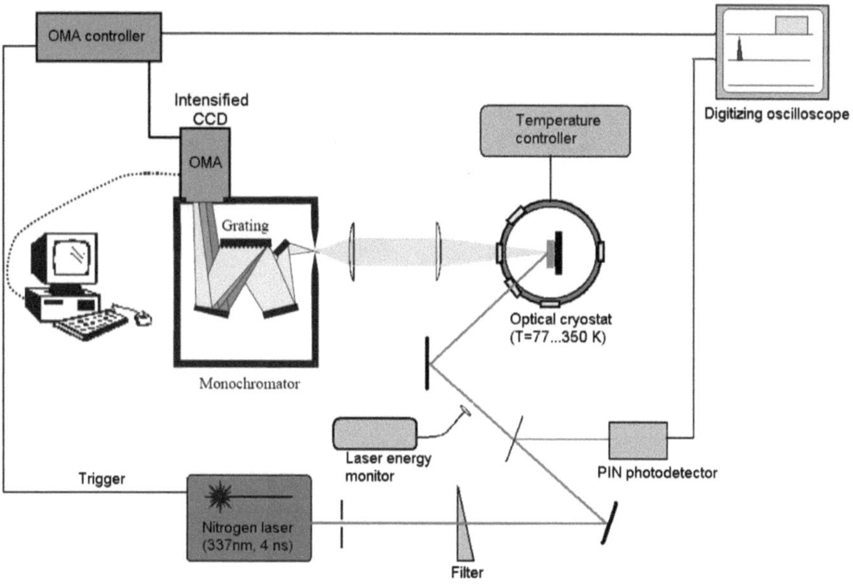

Fig. 2.11 Schematical illustration of the time-resolved photoluminescence measurement setup

2.2.4 Time-Resolved Pump-Probe Experiments

As discussed in Chap. 1 (Sect. 1.3.3) transient absorption measurements or time-resolved pump-probe measurements allow ultra-fast measurements of photo-excitations. With this technique the absorbance at a particular wavelength or range of wavelengths of a sample is measured as a function of time after excitation by a fast femtosecond laser pulse (pump laser). Pump and probe experiments are particularly useful to analyze gain in organic compounds. Indeed, at wavelengths where there is gain, the probe beam will be amplified, whereas at wavelengths where competing absorption processes are dominant, the probe beam is attenuated.

Time resolved pump-probe experiments were performed in close collaboration with RWTH in Aachen. The experimental setup realized at RWTH is illustrated in Fig. 2.12. The key component is a femtosecond regenerative Ti:sapphire amplifier (Spectra-Physics Spitfire XP Shortpulse), which allows excitation fluences up to 3 mJ per pulse and delivers a pulse train at a repetition rate of 1 kHz. At the center wavelength of 800 nm the pulse duration was about 35 fs. A beamsplitter splitted this beam into an intense pump beam for optical excitation and a weak probe beam. For efficient optical excitation of the organic films, an ultrathin nonlinear BBO crystal was inserted into the pump beam to generate optical pulses at 400 nm wavelength via second harmonic generation (SHG). The duration of these 400 nm pulses was ~ 40 fs. Furthermore, the pump beam was chopped mechanically at 500 Hz and excited the sample at $t = 0$. The excitation spot was circular with a full-width-half-maximum (FWHM) of 400 μm (Gaussian profile) and the pump fluence was determined as the peak value at the center of the pump spot. To prevent photo-oxidation

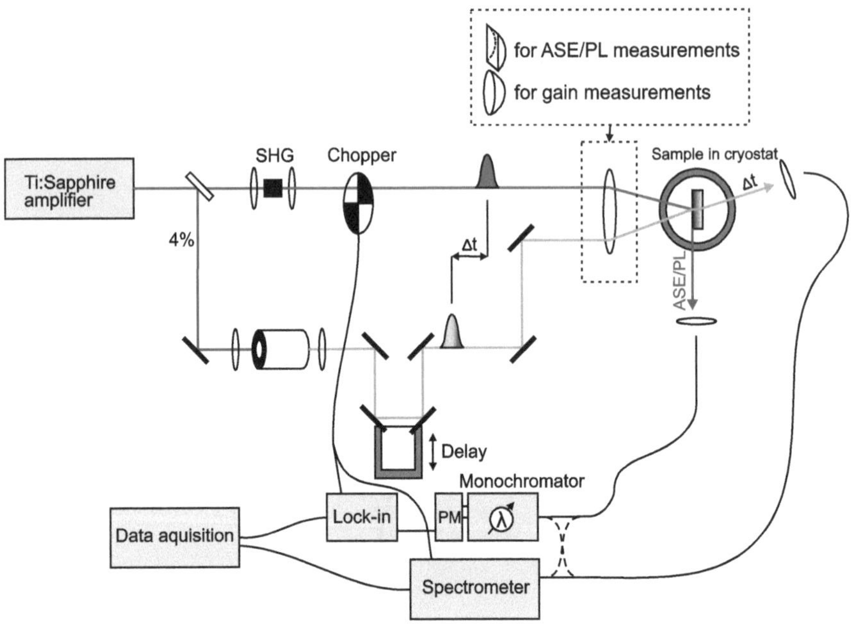

Fig. 2.12 Experimental setup for ASE/PL- and time resoved pump-probe measurements

during the measurements, the samples were mounted in a cryostat. As can be seen on Fig. 2.12, a fraction of the fundamental laser output was spectrally broadened by a sapphire crystal, delayed by a motorized delay stage and focused on the pump illuminated sample spot. This was the probe beam. The much smaller probe spot (50 μm FWHM) was carefully aligned to the center of the pump spot. The transmitted part of the probe beam was guided by an optical fiber to a monochromator and its intensity was measured by a photomultiplier. A lock-in amplifier was synchronized to the chopper frequency and measured the pump induced transmission change. Typically, the experiments were carried out by changing the wavelength for a particular time delay. Afterwards, the time delay was set to the next value and the measurement was repeated.

2.2.5 *Amplified Spontaneous Emission and Loss Measurements*

Amplified spontaneous emission and loss measurements were performed using the same setup as for time-resolved pump-probe experiments realized at RWTH in Aachen (Fig. 2.12). Only some minor modifications were necessary. The measurements were carried out by collecting and detecting the light emitted from the sample in the optically pumped region. The probe beam was blocked in this case. The pump beam ($\lambda = 400$ nm) on the other hand, was elliptically focused on the sample by means of a cylindrical lens, resulting in a narrow excitation stripe. When the

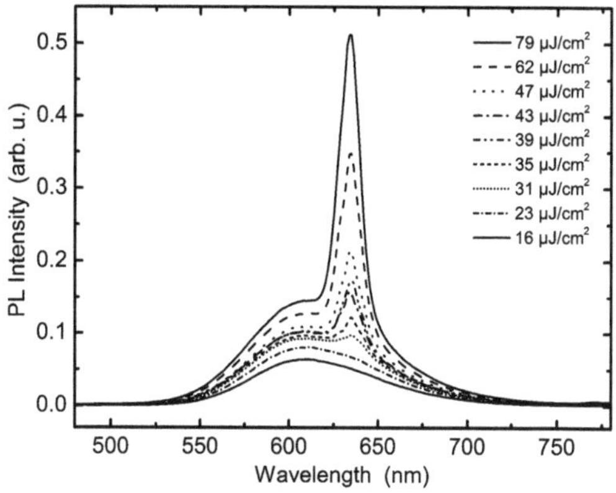

Fig. 2.13 Emission spectra of Alq3:DCM for different laser fluences at 400 nm excitation wavelength (Measurements performed by RWTH Aachen)

pump spot illuminated a region near the edge of the sample, the facet of the organic layer emitted a sufficient amount of light for optical detection. For different average pump powers, the edge-emitted spectrum could then be recorded by guiding the emitted light by an optical fiber to a spectrometer or a monochromator and photomultiplier tube. Similar to the pump-probe measurements, the samples were mounted in a cryostat to prevent degradation and especially photo-oxidation during experiments. Besides, the sample temperature could be controlled in the range from 10 K to room temperature. ASE measurements carried out on a 300-nm thick Alq3:DCM film at a concentration of 2% DCM are shown in Fig. 2.13. The obtained spectra are clearly indicate the dependence of the spectral emission components on the excitation fluence.

Waveguide losses were analyzed by monitoring the edge-emitted light generated in the organic layer by ASE at different distances between the pump spot and the waveguide edge [168]. Since this technique requires a well-defined pump spot, a slit was inserted in the pump beam path to cut off the tails of the elliptical spot in the elongated direction. The pump spot was gradually moved away from the waveguide edge by translating the sample perpendicularly to the laser beam. Since the output intensity from the end of the excitation stripe is assumed to be constant, the observed decreases in the edge-emitted signal result from waveguide losses (mainly absorption and scattering) within the unpumped region. The waveguide loss coefficient α could be determined according to

$$I(z) = I_0 \exp(-\alpha z), \tag{2.7}$$

where I_0 is the light intensity at $z = 0$.

2.2.6 Determination of Thermally Stimulated Luminescence

Thermally stimulated luminescence (TSL) is the phenomenon of lumi nescent emission after removal of excitation under conditions of increasing temperature. Generally, in the TSL method the trapping states are first populated by photogeneration of

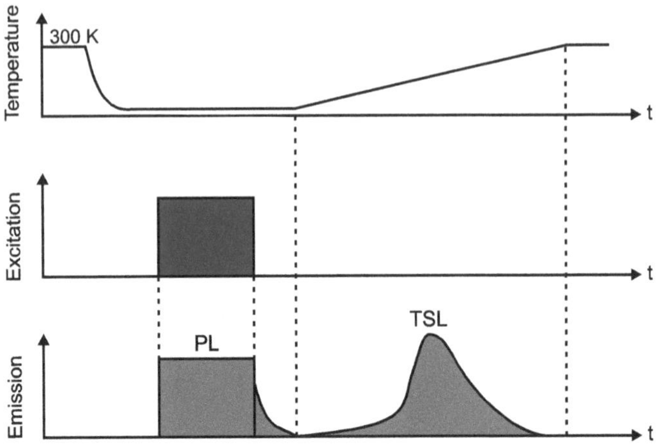

Fig. 2.14 Measuring procedure of thermally stimulated emission experiments

charge carriers, usually at low temperatures in order to prevent a fast escape. Then, the trapped charge carriers are released by heating up the sample with a linear temperature ramp while the luminescence due to radiative recombination is recorded as a function of temperature. This procedure is illustrated in Fig. 2.14. If an energy distribution of the trap states exists, TSL spectra are a complicated convolution of contributions from different traps at different energies and fractional heating techniques have to be applied, which are based on cycling the sample with a large number of small temperature oscillations superimposed on a constant heating run. Consequently, the TSL is a useful tool for determining the trap depths even when traps are not well separated in energy or are continuously distributed. In addition, TSL also allows analysis of the trap spectra even when they are complex.

TSL measurements were carried out at the National Academy of Science of Ukraine using a home-built setup operable from 4 K to 350 K using a temperature controlled helium cryostat. After cooling down to 4 K, the samples were photo-excited, usually for 30 s, by a high-pressure 500 W mercury lamp with an appropriate set of glass optical filters for light selection. After the photo-excitation, the TSL was detected in a photon-counting mode with a cooled photomultiplier, positioned next to the cryostat window. The TSL measurements were performed either at a constant heating rate of 0.15 K/s or in the fractional heating regime. The latter procedure allows the determination of trap depth when different groups of traps are not well separated in energy or are continuously distributed. A more detailed description of the TSL method is described in literature [242, 243].

Chapter 3
Organic Light-Emitting Diodes with Field-Effect Electron Transport

In standard OLEDs the total thickness of the organic layers is generally limited to 80 to 100 nm to obtain efficient charge transport at reasonable driving voltages [244]. Consequently, light is generated very close (within ca. 50 nm) to the metallic cathode and the proximity of the metal electrode induces severe optical absorption losses if the OLED is used as a waveguide. In order to sufficiently reduce these optical losses a possible strategy is to increase the thickness of the organic layers [84, 215]. The low carrier mobilities in organic semiconductors [31, 32], however, prevent an arbitrary increase of this thickness and other techniques, such as the use of transparent conductive oxides [96, 217, 218] and careful optical design [182, 219], are necessary to minimize the losses induced by the electrodes.

In LEOFET devices on the other hand, charge carriers have a field-effect mobility which is typically several orders of magnitude higher than the charge carrier mobility in conventional OLEDs [46], and depending on the type of LEOFET, the light-emission zone can be located at a relatively large distance from the metallic source and drain electrodes [65, 66]. Furthermore, high current densities have been reported for LEOFET devices [63, 69, 75]. However, to obtain these properties, three different electrodes must be used, namely the source, the drain and the gate. In addition, most organic field-effect materials demonstrate rather weak photoluminescence and vice versa, most organic light-emitting materials have poor field-effect performance [57]. Indeed, a high carrier mobility, as needed in transistors, is favored by a tight intermolecular π stacking. Examples are pentacene [245] and region-regular P3HT [246]. However, the photoluminescence yield is usually low in such materials with strong intermolecular coupling. Conversely, materials with high photoluminescence quantum yield, e.g. PPVs [247] are usually characterized by a low intermolecular coupling, and thus limited hopping transport between molecules [248]. Therefore, it would be desirable to form a more elaborated heterojunction consisting of charge transport layers and a light-emitting layer, which can be optimized separately [249]. Such heterojunction concepts are commonly used in high-performance OLEDs [33, 34]. In LEOFET device architectures, however, this approach is difficult to apply.

In this chapter, a novel two-electrode light-emitting device structure is proposed. The device is a hybrid structure between a diode and a field-effect transistor. Com-

S. Schols, *Device Architecture and Materials for Organic Light-Emitting Devices*,
DOI 10.1007/978-94-007-1608-7_3,

pared to conventional OLEDs, the cathode is displaced one to several micrometers from the light-emission zone. As the light emission zone is not covered by metal, the device can be used for top emission or even as waveguide. The micrometer-sized distance between the cathode and the active region can be bridged by electrons with an enhanced field-effect mobility. Owing to this high charge carrier mobility, large current densities are possible. The external quantum efficiency at these high current densities is as high as that of conventional OLEDs using the same materials. In contrast to organic light-emitting field-effect transistors, only two electrodes are used. Moreover, light-emission in the novel device structure always occurs at a fixed position, irrespective of the applied bias, in contrast to LEOFETs where the emission zone can move within the channel by varying the bias conditions [65–67]. The first section of this chapter describes the technology and the materials used to fabricate the device. Next, the device operation and the device performance are discussed. The electrical characteristics as well as the opto-electronic performance are studied. Suggestions for further improvement of the device performance are given in the last section of this chapter. The content of this chapter is published in [250, 251].

3.1 Device Fabrication

The schematic architecture of the device is illustrated in Fig. 3.1. The device is fabricated on top of an ITO-coated glass substrate and comprises an organic hole-transporting layer (HTL), an organic light-emitting layer and an organic electron-transporting layer (ETL). Prior to the deposition of these organic layers, an insulating layer of 100 nm SiO_2 is deposited by sputtering. Trenches are selectively wet-etched through the SiO_2, to be able to contact the ITO layer. ITO serves as the hole-injecting electrode in the device. The cathode is formed by a thin layer of 0.8 nm lithium fluoride (LiF) followed by the deposition of 100 nm aluminum (Al).

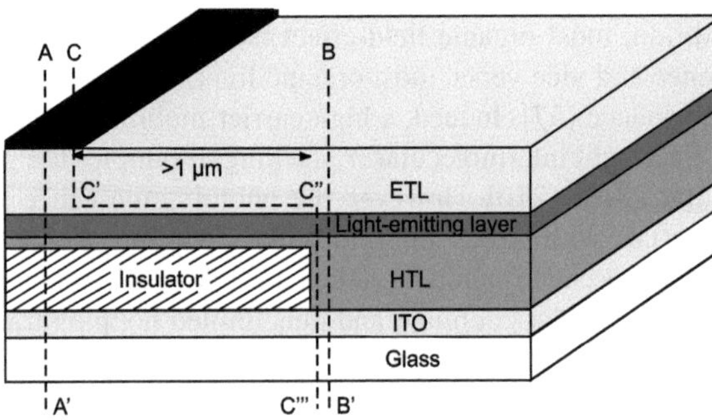

Fig. 3.1 Schematic architecture of the organic light-emitting diode with field-effect electron transport. The device comprises an organic hole-transporting layer (HTL), an organic electron-transporting layer (ETL) and an organic light-emitting layer. The dashed lines AA′, BB′ and CC′C″C‴ indicate three different sections. (© Wiley-VCH Verlag GmbH & Co. KGaA. Reproduced with permission)

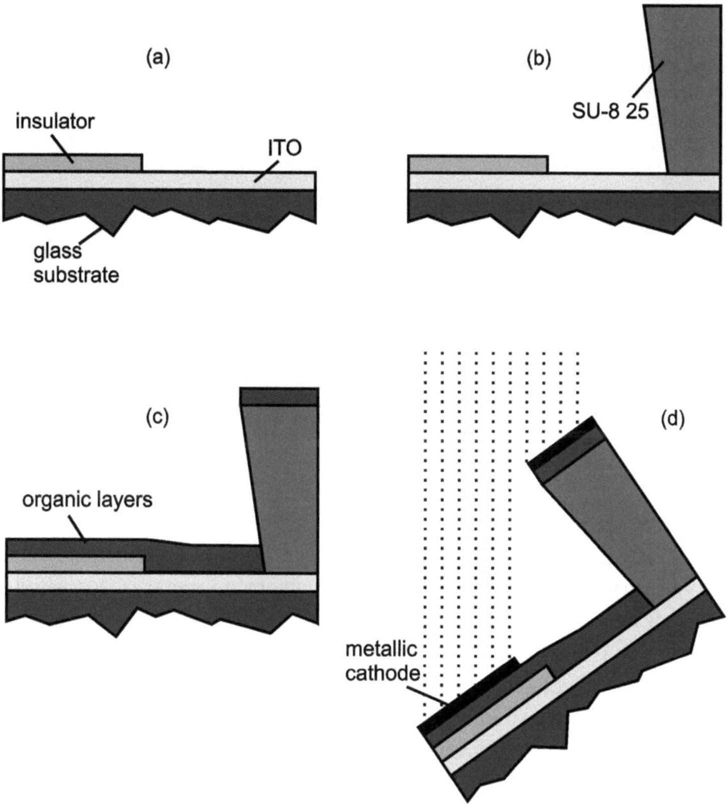

Fig. 3.2 Schematic depiction of the processing sequence for the fabrication of the light-emitting field-effect device structure: (**a**) situation before creation of the integrated shadow mask, (**b**) deposition of the integrated shadow mask, (**c**) deposition of the organic layers, and (**d**) deposition of the metallic cathode with the substrate mounted on a triangular sample holder. During this deposition process, the atomic flux is at a 45° angle with respect to the substrate. (© Wiley-VCH Verlag GmbH & Co. KGaA. Reproduced with permission)

This cathode is not positioned vertically above the ITO anode in the trench in the SiO_2 insulator, but it is located fully above the insulator, as shown in Fig. 3.1. The distance between the metallic top electrode and the insulator edge is one to several microns. Accurate alignment of the metallic cathode is obtained by using an integrated shadow mask technique [252] and angled deposition of the LiF/Al layer (Fig. 3.2).

The integrated shadow mask is realized by patterning a 20-μm thick negative photoresist SU-8 25 (MicroChem Corp., used as received). This layer is resistant to solvents, acids and bases and has excellent thermal stability. The different organic layers are deposited after application and patterning of the SU-8 25 resist. Subsequently, the sample is mounted on a triangular sample holder and loaded in the ultra-high vacuum system for cathode evaporation. During this deposition, the flux is at a 45° angle with respect to the substrate. The SU-8 25 profile thus creates

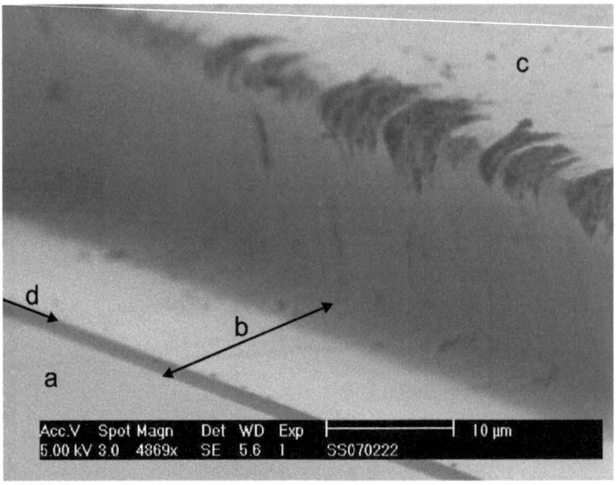

Fig. 3.3 SEM image of the device structure: (**a**) metallic cathode, (**b**) shadowed region where no metal is deposited, (**c**) the 20-μm thick SU-8 25 layer and (**d**) arrow indicating the insulator edge. (© Wiley-VCH Verlag GmbH & Co. KGaA. Reproduced with permission)

a shadowed region with a span similar to the thickness of the SU-8 25, so that the substrate is only partially covered with the metal. Figure 3.3 shows a SEM image of the device structure. The shadowed region where no metal is deposited can be recognized. It can also be verified that the walls of the SU-8 25 layer are slightly re-entrant, which is typical for negative photoresists.

PTAA [253] is selected as the hole-transporting material to fabricate the device. This material can be deposited by spincoating, which results in a film with a smooth top surface, favorable for the growth of an additional layer. The bulk hole mobility of PTAA is approximately 10^{-2} cm^2/Vs [253]. For the active light-emitting layer 20 nm Alq_3 doped with 2% of DCM_2 is chosen. DCM_2 is a well-known red light-emitting fluorescent dye used in OLEDs [254], with a photoluminescence quantum yield of about 40% [41]. The organic layer stack is finalized by the deposition of an organic electron-transporting layer. This electron-transporting material is selected according to two criteria which are important in our device architecture. First, the organic electron-transporting material should have a high electron field-effect mobility since this mobility determines the performance. Numerical simulations indicate that the electron mobility is much more critical for the device performance than the mobility of the holes. Second, as electron transport is intended to occur at the heterojunction between the electron-transporting layer and the light-emitting layer, the LUMO of the electron-transporting material should be slightly lower than the LUMO of Alq_3 and DCM_2. PTCDI-$C_{13}H_{27}$ satisfies these two conditions and therefore it is used as the electron-transporting material. Typically, 50 nm PTCDI-$C_{13}H_{27}$ is deposited at a deposition rate of 0.5 Å/s, while the substrate is kept at room temperature. Gundlach *et al.* reported a field-effect mobility of about 0.6 cm^2/Vs for PTCDI-$C_{13}H_{27}$ using chromium (Cr) top-contacts [255]. A mobility of 0.28 cm^2/Vs was obtained using LiF/Al top-contacts. Even higher field-effect mobilities can be achieved by annealing after fabrication [256]. The molecular structures and the energy levels of the HOMO and the LUMO of PTAA, Alq_3, DCM_2 and PTCDI-$C_{13}H_{27}$ can be found in Chap. 2 (Sect. 2.1.1).

3.2 Device Operation

To explain the device operation we draw in Fig. 3.4 the band diagrams through sections AA′, BB′ and CC′C″C‴ of Fig. 3.1 for the device in forward bias, i.e. for a positive anode-to-cathode bias. Under these conditions electrons are injected from the cathode into PTCDI-$C_{13}H_{27}$ at cross-section AA′. The semiconductor PTAA and the light-emitting layer Alq_3:DCM_2 on top of the SiO_2 (cross-section AA′) are depleted of holes. As will be confirmed further, the LUMO offset at the interface between Alq_3 and PTCDI-$C_{13}H_{27}$ prohibits electron injection from PTCDI-$C_{13}H_{27}$ into Alq_3 in the cross-section AA′. As a result, the depleted PTAA layer and Alq_3:DCM_2 layer on top of the SiO_2 insulator behave as a dielectric layer in series to the SiO_2 insulator, and an electron accumulation layer is formed in the PTCDI-$C_{13}H_{27}$ at the interface with Alq_3:DCM_2. These electrons are transported laterally by the electric field from the cathode towards the SiO_2 edge (from C′ to C″). The field-effect mobility of these charge carriers is larger than the charge carrier mobility in a conventional OLED, making transport over several microns possible. Near the insulator edge, at position C″, electrons are injected into the Alq_3:DCM_2 layer, where they recombine with holes, injected from the ITO and transported vertically through the PTAA (see cross-section BB′ in Fig. 3.4). The electron injection at position C″ into Alq_3 is enabled by an enhanced vertical electric field at that position.

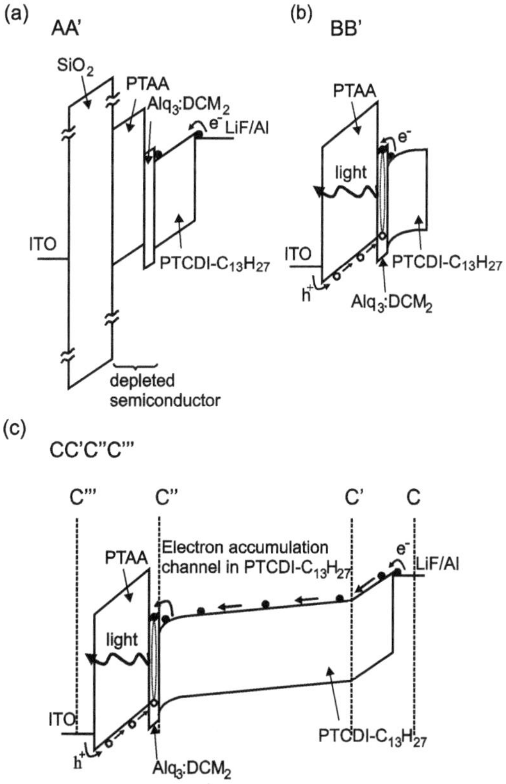

Fig. 3.4 Band diagrams for the device under forward bias conditions through different sections of Fig. 3.1: (**a**) cross-section AA′, (**b**) cross-section BB′ and (**c**) section CC′C″C‴. (© Wiley-VCH Verlag GmbH & Co. KGaA. Reproduced with permission)

The enhancement of the vertical field at C″ has two causes: first, the vertical and lateral electric fields in the channel (section C′C″) collapse into only a vertical field at position C″; furthermore, the presence of a space charge of holes in PTAA increases the vertical electrical field that attracts electrons to be injected in the Alq_3:DCM_2 layer. This latter behavior is similar to the enhanced hole injection as a result of a space charge of electrons at the anode of OLEDs reported by Van Woudenbergh *et al.* [257]. Exciton formation occurs and radiative relaxation of these excitons to the ground state results in light-emission near the SiO_2 edge, at a micrometer-size distance from the metallic contact. In this way, this heterojunction device allows to minimize optical losses at the metal cathode. The vicinity of the ITO bottom-contact is of minor importance with respect to absorption losses because it has been shown that thin transparent ITO layers have low optical losses in the 550–750 nm spectral range [184]. Note that since the potential near the insulator edge is floating and since there is no electron confinement in the lateral direction (the electrons are not hindered by an interface), the electron and hole current in the device are always balanced.

We have confirmed this view of the device operation by measuring the characteristics of several control devices. Control transistors comprising the stacks PTAA/PTCDI-$C_{13}H_{27}$ and PTAA/Alq_3:DCM_2/PTCDI-$C_{13}H_{27}$ have been fabricated on doped n^{++}-Si wafers with a 100-nm thick thermally grown SiO_2 layer (serving as the OTFT gate dielectric). LiF/Al was used as source and drain top electrodes. For the PTAA/PTCDI-$C_{13}H_{27}$ transistor, a field-effect mobility for electrons in PTCDI-$C_{13}H_{27}$ at the heterojunction interface with PTAA of 0.2 cm^2/Vs is extracted. It is striking that the hole-transporting organic semiconductor PTAA provides a good growth surface for the electron-conducting PTCDI-$C_{13}H_{27}$, and that the hetero-interface is appropriate for electron transport. This can be attributed to the smooth top surface of PTAA and the formation of a high-quality interface free of electron traps. We infer the latter from the observation that the electron mobility in PTCDI-$C_{13}H_{27}$ transistors on PTAA-coated SiO_2 is the same as the one we measured in control transistors that have a gate dielectric consisting of SiO_2 coated with PαMS, known to provide a high-quality, electron-trap free surface [258] that allows good electron transport [259].

Upon inserting Alq_3:DCM_2 between PTAA and PTCDI-$C_{13}H_{27}$, the effective saturation field-effect electron mobility drops to 0.08 cm^2/Vs. The output and transfer characteristics of this transistor are shown in Fig. 3.5. We correlate the reduction in mobility to the different growth of PTCDI-$C_{13}H_{27}$ on Alq_3:DCM_2 as compared to growth on PTAA. Figures 3.6(a) and (b) show the PTCDI-$C_{13}H_{27}$ morphology on, respectively, PTAA and PTAA/Alq_3:DCM_2. PTCDI-$C_{13}H_{27}$ deposited on PTAA/Alq_3:DCM_2 reveals a much rougher growth than PTCDI-$C_{13}H_{27}$ grown on top of PTAA; this correlates with the lower field-effect mobility. The rms roughness of the layer is increased from 6.7 nm to 10.7 nm upon inserting an Alq_3:DCM_2 layer. As both substrates (PTAA and PTAA/Alq_3:DCM_2) are morphologically similar, being amorphous flat surfaces with the same rms roughness of 0.4 nm, the difference in growth of PTCDI-$C_{13}H_{27}$ originates from the interfacial energy differences between, respectively, PTCDI-$C_{13}H_{27}$ and PTAA and PTCDI-$C_{13}H_{27}$ and Alq_3:DCM_2.

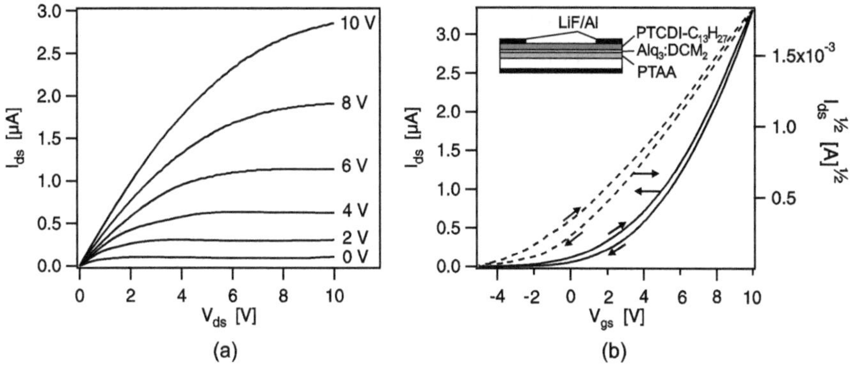

Fig. 3.5 Electrical characteristics of a top-contact transistor comprising PTAA/Alq_3:DCM_2/PTCDI-$C_{13}H_{27}$ as the organic layers ($W/L = 2000/50$). (**a**) Output characteristics for various V_{gs}, and (**b**) transfer characteristics of the same transistor. A saturation mobility of 0.08 cm^2/Vs and a threshold voltage of ~ 0 V are extracted for this device. (© Wiley-VCH Verlag GmbH & Co. KGaA. Reproduced with permission)

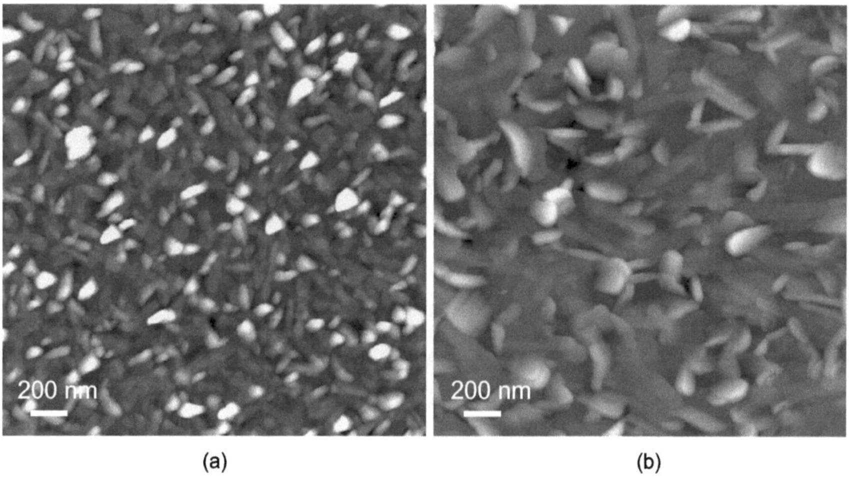

Fig. 3.6 2 µm × 2 µm atomic force microscopy (AFM) surface scans of a 50 nm thick PTCDI-$C_{13}H_{27}$ layer on top of (**a**) PTAA and (**b**) PTAA/Alq_3:DCM_2. On PTAA/Alq_3:DCM_2 a higher surface roughness is measured: the rms roughness is 10.7 nm for PTCDI-$C_{13}H_{27}$ on PTAA/Alq_3:DCM_2 versus 6.7 nm for PTCDI-$C_{13}H_{27}$ grown on PTAA. (© Wiley-VCH Verlag GmbH & Co. KGaA. Reproduced with permission)

Additionally, a control transistor was fabricated comprising PTAA/Alq_3:DCM_2 as the organic layers, to verify whether Alq_3:DCM_2 could not be used as the electron-transport layer instead of PTCDI-$C_{13}H_{27}$. No field-effect transport of electrons could be observed in this structure. The non-planar molecular structure of Alq_3 apparently prohibits efficient lateral electron transport. This control experiment also proves that electron transport in the transistor comprising the organic stack PTAA/Alq3:DCM_2/PTCDI-$C_{13}H_{27}$ and in our novel device structure (between position C′ and C″ in Fig. 3.1) indeed occurs in the PTCDI-$C_{13}H_{27}$ and not in Alq3:DCM_2.

3.3 Device Performance

3.3.1 Optical and Electrical Characterization

Figure 3.7(a) displays the experimentally measured electrical characteristics of a light-emitting device constructed by using the architecture depicted in Fig. 3.1. The device has a width of 1 mm and was measured in an inert N_2 atmosphere immediately after evaporation. The metallic cathode of the device is displaced by 3.5 μm with respect to the insulator edge. Under forward bias, the current increases with increasing voltage. The electrical characteristics can be explained by the equivalent

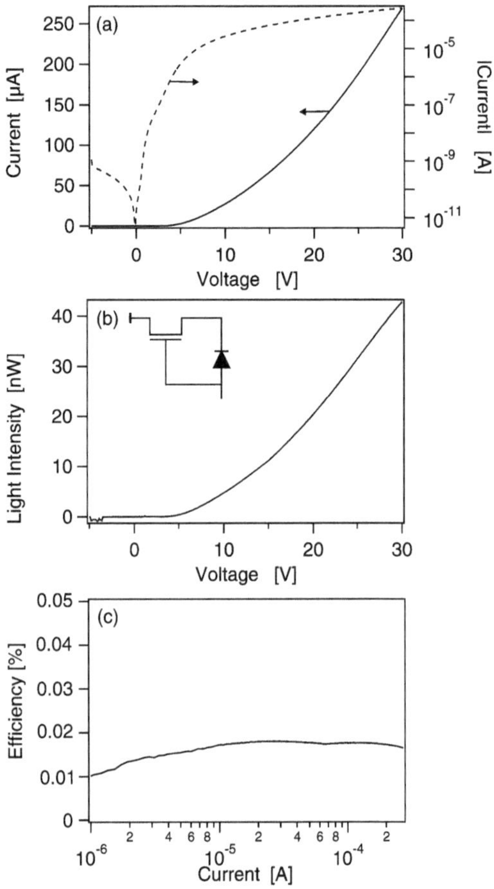

Fig. 3.7 Experimentally measured characteristics of the light-emitting diode with field-effect electron transport. This device has a width of 1 mm and a distance between the top electrode and the insulator edge of 3.5 μm. (**a**) Current-voltage characteristics, (**b**) corresponding light output. *Inset*: equivalent circuit of the device, and (**c**) external quantum efficiency as a function of the current. (© Wiley-VCH Verlag GmbH & Co. KGaA. Reproduced with permission)

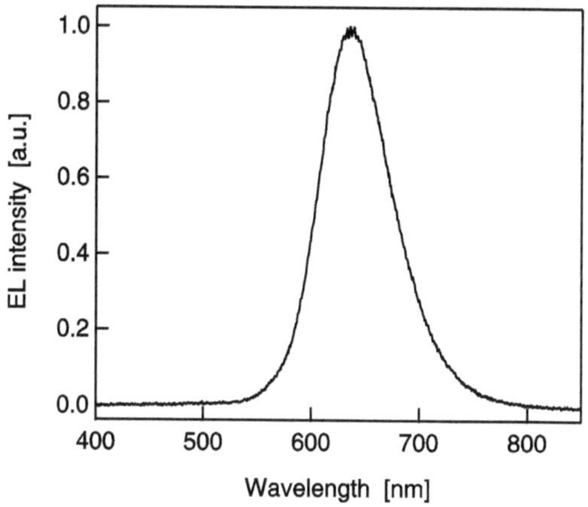

Fig. 3.8 Normalized electroluminescence spectrum of the light-emitting diode with field-effect electron transport. The spectrum corresponds to DCM_2 emission with a maximum positioned at wavelength $\lambda = 636$ nm. (© Wiley-VCH Verlag GmbH & Co. KGaA. Reproduced with permission)

circuit shown as inset in Fig. 3.7(b). This equivalent circuit consists of an n-type transistor having a diode between the gate and the drain.

The device emits red light upon the radiative decay of excitons. The optical output intensity as a function of the applied voltage bias is shown in Fig. 3.7(b). It can be seen that the optical output tracks the current characteristics. The normalized electroluminescence (EL) spectrum of the device is shown in Fig. 3.8. The peak emission wavelength is located at 636 nm and corresponds to DCM_2 emission. Two control experiments allowed to verify that the observed emission indeed originates from DCM_2. In a first control experiment, we changed the dye from DCM_2 to $Btp_2Ir(acac)$ [260], a phosphorescent dye with characteristic spectral features that could easily be recognized. In another control experiment, we used undoped Alq_3; this device emits green light, peaking at 540 nm, as is characteristic for emission from Alq_3.

In our device structure, the emission zone is well-defined and light-emission always occurs at a fixed position, namely near the edge of the insulator. This position is independent of the applied bias. The left panel of Fig. 3.9 shows a photograph in reflection of a device without biasing. The metal electrode (white reflecting area) and the insulator edge, indicated with an arrow, can be recognized. For this device, the distance between the cathode and the insulator edge is 19 μm. The right panel of Fig. 3.9 is an image in the dark of the optical output when the device is in forward bias. A narrow line of red light appears alongside the edge of the insulator. The light intensity increases with the bias voltage. Despite the narrow emission zone, the red light can easily be observed by the naked eye. The measured width of the emission zone is about 2 μm.

3.3.2 Analysis

The experimentally observed behavior is qualitatively confirmed by numerical simulations. A simplified device, comprising a hole-transporting layer (HTL) and a

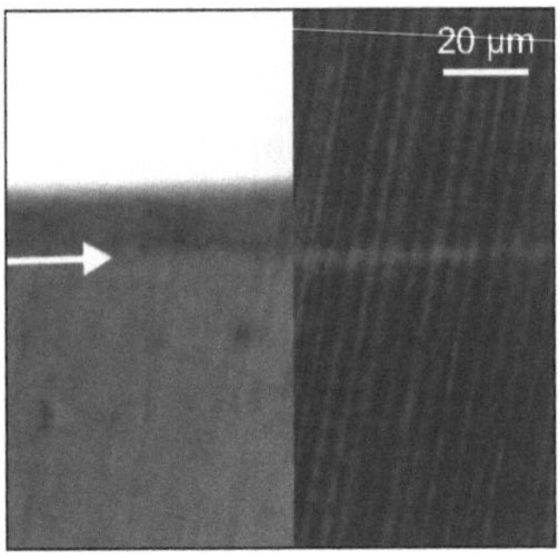

Fig. 3.9 *Left panel*: Optical microscope reflection image of a device without bias. The *white area* is the reflective metal cathode, the *arrow* indicates the insulator edge. *Right panel*: under forward bias a narrow line of light appears along the insulator edge. The width of the line is estimated to be about 2 μm. (© Wiley-VCH Verlag GmbH & Co. KGaA. Reproduced with permission)

light-emitting electron-transpor ting layer (ETL), was simulated using the Silvaco ATLAS 2D device simulator. The material parameters used in the simulations are summarized in Table 3.1. For the ETL an electron mobility of $8 \cdot 10^{-2}$ cm^2/Vs is used, which corresponds to the experimentally measured electron field-effect mobility of PTCDI-$C_{13}H_{27}$ in our device. The hole mobility of the ETL is taken to be $5 \cdot 10^{-4}$ cm^2/Vs. For the HTL on the other hand a hole mobility of $5 \cdot 10^{-3}$ cm^2/Vs and a negligible electron mobility are taken. The HOMO and LUMO of the ETL and HTL are assumed to be similar to the HOMO and LUMO of, respectively, PTCDI-$C_{13}H_{27}$ and PTAA. Typical values for the singlet lifetime (τ) and the exciton diffusion length (L_{diff}) are taken from literature [261]. The top contact is assumed to be located at 6 μm from the insulator edge. Figure 3.10(a) illustrates the simulated recombination zone as the result of applying a positive bias to the anode. It can be clearly seen that recombination occurs near the insulator edge, several microns from the metallic contact. The simulated recombination zone has a width of about 2 μm, which is in agreement with the experimental observation. Figure 3.10(b) confirms the presence of an electron accumulation layer at the interface between the HTL and the light-emitting ETL when a positive bias is applied to the anode with respect to the cathode. This electron accumulation region vanishes beyond the insulator edge, as electrons and holes recombine there.

Table 3.1 Summary of the parameters used in the 2D device simulator

	HTL	Light-emitting ETL
HOMO (eV)	5.1	5.4
LUMO (eV)	1.8	3.4
μ_e (cm^2/Vs)	$1 \cdot 10^{-7}$	$8 \cdot 10^{-2}$
μ_h (cm^2/Vs)	$5 \cdot 10^{-3}$	$5 \cdot 10^{-4}$
L_{diff} (nm)		10
τ (s)		$16 \cdot 10^{-9}$

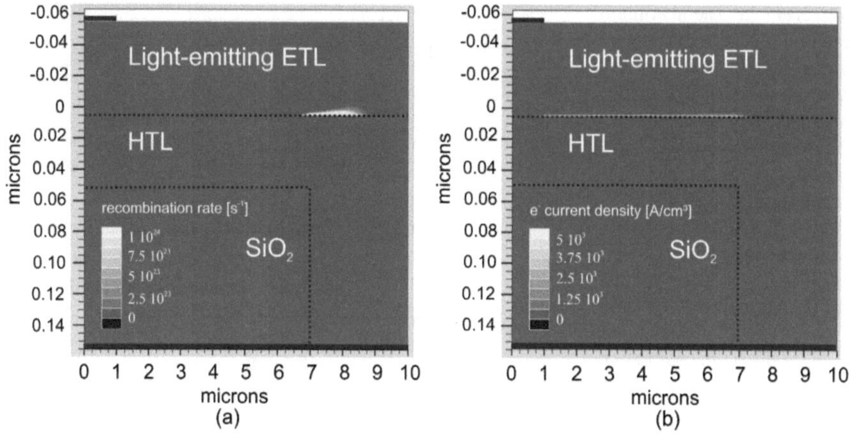

Fig. 3.10 (**a**) The simulated recombination rate, and (**b**) the simulated electron accumulation layer under forward bias. (© Wiley-VCH Verlag GmbH & Co. KGaA. Reproduced with permission)

The external quantum efficiency η_{ext} of the fabricated devices is estimated based on luminance, EL spectra and current. The measured maximum η_{ext} of the device is 0.02%, which is similar to that of a light-emitting diode using the same materials. Reference OLEDs with Alq_3:DCM_2 active layers have been shown with up to 0.5% efficiency [254]. In these reference OLEDs, Alq_3 is used as electron transporting layer, rather than PTCDI-$C_{13}H_{27}$. However, Alq_3 cannot be used in our structure because its non-planar molecular structure prohibits efficient electron transport in a thin-film transistor configuration.

By changing the distance between the metallic contact and the insulator edge, the light intensity of the OLED with field-effect electron transport can be tuned. We measured different devices for which the top-contact displacement with respect to the insulator edge is systematically varied from 0.8 μm to 9 μm. In Fig. 3.11(a) the measured maximum η_{ext} of each of these devices is plotted as a function of the distance between the cathode and the insulator edge. As expected, the maximum η_{ext} is constant.

On the other hand, at the average breakdown voltage of the SiO_2 insulator, about 30 V, higher currents are possible in devices with shorter cathode displacement with respect to the insulator edge. As a result, the maximum achievable optical output power and brightness can be higher for smaller distances. This is illustrated in Fig. 3.11(b) which shows the light intensity of three different devices with distances between the metallic cathode and the insulator edge of, respectively, 7.3 μm, 5.1 μm and 3.5 μm.

Taking into account a 2 μm wide emission zone, the maximum hole current density in the device at 30 V is estimated to be 13 A/cm^2. This current density is much higher than the current density obtained in conventional OLEDs (typically in the order of 10^{-2} A/cm^2 at the point of maximum η_{ext}). The maximum electron current density in the accumulation layer of the light-emitting device is even higher. Assuming a current flow confined to a 5-nm thick electron accumulation layer, which

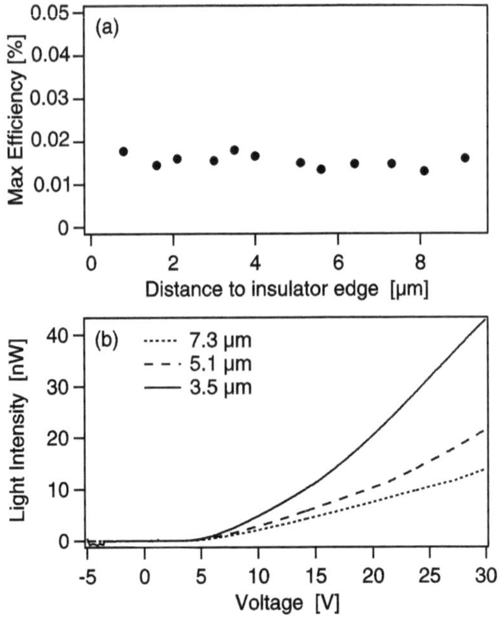

Fig. 3.11 (**a**) The maximum external quantum efficiency η_{ext} as a function of the distance to the insulator edge. Irrespective of the distance, the maximum η_{ext} is constant. (**b**) Light intensity versus applied voltage of three different devices for which the top-contact displacement with respect to the insulator edge is 7.3 μm (*dotted lines*), 5.1 μm (*dashed lines*) and 3.5 μm (*full lines*), respectively. (© Wiley-VCH Verlag GmbH & Co. KGaA. Reproduced with permission)

corresponds to approximately two PTCDI-$C_{13}H_{27}$ monolayers [256], current densities of 5400 A/cm^2 are achieved. This is comparable to the highest current densities achieved in ambipolar organic light-emitting transistors reported to date [63, 75] and can be attributed to the high electron mobility in our device. Figure 3.7(c) shows the measured η_{ext} as a function of the current. Within the experimental error no significant roll-off is observed up to the maximum current.

3.4 Improvement of the External Quantum Efficiency

The measured maximum η_{ext} of the fabricated device is 0.02%. This value is rather low compared to reference OLEDs based on Alq_3:DCM_2 reported in literature [254]. However, in these reference OLEDs, Alq_3 is used as the electron-transporting material, instead of PTCDI-$C_{13}H_{27}$, indicating that the presence of PTCDI-$C_{13}H_{27}$ might be responsible for the low η_{ext} in our devices.

The low η_{ext} of the devices may have two plausible origins: (i) exciton dissociation at the interface between the Alq_3:DCM_2 and the PTCDI-$C_{13}H_{27}$ layer; or (ii) re-absorption of the emitted photons by one of the used organic materials. The first of the above mentioned mechanisms would mean that excitons of DCM_2 (with LUMO of 3.1 eV and HOMO of 5.2 eV) [235] would dissociate to PTCDI-$C_{13}H_{27}$ (with LUMO of 3.4 eV) [236] by electron-transfer at the interface between these two materials. To test whether this mechanism is operative, we fabricated devices in which a thin undoped layer of Alq_3 was inserted between the light-emitting layer and the electron-transporting layer. Such a layer would prohibit the dissociation of

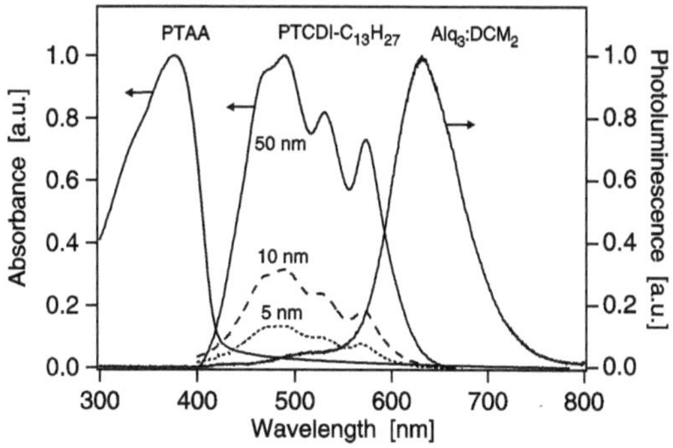

Fig. 3.12 Normalized photoluminescence spectrum of a thin film of Alq_3 doped with 2% of DCM_2 and absorption spectra of PTAA and PTCDI-$C_{13}H_{27}$ thin films with different thicknesses: 50 nm (*full line*), 10 nm (*dashed line*) and 5 nm (*dotted line*). (© Wiley-VCH Verlag GmbH & Co. KGaA. Reproduced with permission)

excitons of DCM_2 because the LUMO of Alq_3 (2.8 eV) [235] is higher than the LUMO of DCM_2. Therefore, if the low η_{ext} is a result of exciton break-up, it would be expected that higher quantum efficiencies could be measured for diodes with an interfacial layer of undoped Alq_3. Experimentally, we used an interfacial layer of 10 nm of Alq_3. The η_{ext} remained unaltered. Hence, exciton dissociation cannot be identified as the origin for the low η_{ext} in PTCDI-$C_{13}H_{27}$-based devices. Note that an incomplete recombination of holes and electrons due to an imbalance between the electron and hole current can also not be invoked as the reason for the rather low external quantum efficiency, since the electron and hole current in the device are always balanced.

Another possibility is that photons emitted by DCM_2 are re-absorbed by one of the other organic materials that are used to fabricate the device. In Fig. 3.12, the normalized absorption spectra of PTAA and PTCDI-$C_{13}H_{27}$ and the normalized photoluminescence spectrum of Alq_3:DCM_2 are shown. The absorption spectrum of PTAA is clearly blue-shifted with respect to the emission of DCM_2. The emission spectrum of Alq_3:DCM_2 and the absorption spectrum of PTCDI-$C_{13}H_{27}$ on the other hand, display a significant overlap, indicating that photons emitted by DCM_2 will be strongly re-absorbed by PTCDI-$C_{13}H_{27}$. A simple way to reduce this overlap is by decreasing the PTCDI-$C_{13}H_{27}$ layer thickness. In Fig. 3.12 the absorption spectra of very thin (5 nm and 10 nm) PTCDI-$C_{13}H_{27}$ layers are also plotted. These spectra are scaled with the same factor used for the normalized 50-nm thick PTCDI-$C_{13}H_{27}$ film spectrum in order to preserve the relative absorbance between the different samples. From Fig. 3.12 it is clear that the overlap with the Alq_3:DCM_2 light-emitting layer is significantly reduced when thin layers of PTCDI-$C_{13}H_{27}$ are used, indicating a reduction of photon re-absorption.

To confirm this hypothesis, we have fabricated and characterized an OLED with field-effect-assisted electron transport comprising an ultra-thin 10-nm electron-transporting PTCDI-$C_{13}H_{27}$ layer. Characterization was performed in an inert N_2 atmosphere immediately after evaporation of the metallic top contact. The experimentally measured characteristics are shown in Fig. 3.13. The device had a width of 1 mm and the metallic cathode was displaced by 1.7 μm with respect to the insulator edge. An η_{ext} of 0.05% was calculated. This is more than two times larger than the value obtained when 50 nm PTCDI-$C_{13}H_{27}$ was used as the electron-transporting

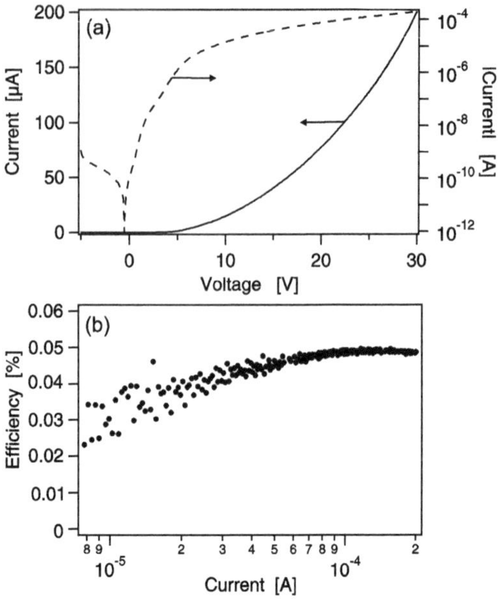

Fig. 3.13 Experimentally measured characteristics of a 1 mm wide device, having a distance of 1.7 μm between the metallic top contact and the insulator edge. In this device 10 nm of PTCDI-$C_{13}H_{27}$was used as the electron-transporting layer. (**a**) Current-voltage characteristics on a logarithmic scale (*dashed line*) and a linear scale (*full line*), and (**b**) η_{ext} as a function of the current. (© Wiley-VCH Verlag GmbH & Co. KGaA. Reproduced with permission)

material. From these measurements we infer that re-absorption of emitted photons by PTCDI-$C_{13}H_{27}$ is indeed limiting the device performance.

The results above clearly indicate that in order to further improve the device performance other electron-transporting materials should be used. Apart from a high electron field-effect mobility and a slightly lower LUMO than the LUMO of Alq_3 and DCM_2, the material should also have low absorption in the device light-emitting wavelengths, in this case the red spectral region.

3.5 Summary and Conclusions

In this chapter, an organic light-emitting diode with field-effect electron transport was proposed as a new device concept complementary to the list of existing electro-luminescent devices. In this device configuration, the metallic top contact is remote from the light-emission zone. A micrometer-sized distance between the cathode and the light-emission zone is bridged by electrons with an enhanced field-effect mobility. The light-emission occurs at a fixed position irrespective of the applied bias. We fabricated the device using Alq_3:DCM_2 as host-guest light-emitting material system, PTCDI-$C_{13}H_{27}$ as electron-transporting material and PTAA as hole-transporting material. Light-emission was found to correlate with the current, and could be modulated by the anode-to-cathode voltage. The measured device operation was confirmed by 2D numerical simulations and by the characteristics of a number of control devices. Devices with smaller distance between the cathode and the insulator edge allowed larger currents at the maximum operating voltage, and therefore also higher brightness. η_{ext} was confirmed to be as high as in a conventional OLED using the same materials. An increase of η_{ext} could be obtained by decreasing the PTCDI-$C_{13}H_{27}$ layer thickness, indicating that re-absorption of emitted photons by PTCDI-$C_{13}H_{27}$ is limiting the device performance. Additional optimization of material combinations may further increase η_{ext}. The quantum efficiency was remarkably constant up to the maximum current, which corresponds to a hole current density in the order of 10 A/cm^2. This high current density, in combination with reduced optical absorption losses thanks to the remoteness of the metal cathode, may lead to interesting applications as waveguide OLEDs and possibly a laser structure.

Chapter 4
Devices Based on Diperfluorohexyl-quaterthiophene Derivatives

For high-brightness OLEDs, and for electrically pumped organic lasers, it is important to implement materials and device configurations that allow high electroluminescence efficiencies at high current densities. In conventional OLEDs high quantum efficiencies have been reported [38, 39, 262–265], even approaching an internal quantum efficiency of 100% [34]. However, in these devices the point of maximum η_{ext} is typically reached at low current densities (in the order of 10^{-2} A/cm^2) because of increased non-radiative losses at higher voltage bias [217, 266–268].

In ambipolar polymer light-emitting transistors on the other hand, current densities of 50 A/cm^2 could be achieved without reduction of η_{ext} ($\sim$0.75%) [67]. This current density could be further increased by using another gate-dielectric [69], although in this case a decrease of η_{ext} was observed at higher bias, which was attributed to singlet-polaron quenching. In addition, Takenobu *et al.* reported current densities of several hundreds A/cm^2 in tetracene and rubrene single-crystal light-emitting transistors [75]. Despite the absence of roll-off the η_{ext} of these single-crystal devices was, however, only on the order of 0.03%. More recently, current densities of kA/cm^2 were reported for solution processed bilayer LEOFETs having an η_{ext} of 0.15% [63].

In our OLEDs with field-effect electron transport a similar behavior has been observed. In the previous chapter (Chap. 3) we have demonstrated that in a device comprising PTAA as the hole-transporting material, Alq_3 doped with DCM_2 as the active light-emitting layer and PTCDI-$C_{13}H_{27}$ as the electron-transporting material, an electron current density of $\sim$ 5 kA/cm^2 could be achieved without significant roll-off of η_{ext}. The maximum η_{ext} of these OLEDs with field-effect-assisted electron transport was however rather low. In a device comprising a 50-nm thick PTCDI-$C_{13}H_{27}$ layer, a maximum η_{ext} of 0.02% could be achieved. This efficiency could be increased to 0.05% by reducing the PTCDI-$C_{13}H_{27}$ layer thickness, indicating that re-absorption of emitted photons by PTCDI-$C_{13}H_{27}$ is a main limitation with respect to device performance. In order to further improve device performance, other materials or material combinations should be used.

In this chapter, two alternative electron-transporting materials are investigated for use as the electron-transporting material in OLEDs with field-effect electron transport: DFH-4T [269] and DFHCO-4T [237]. Both materials were selected according

S. Schols, *Device Architecture and Materials for Organic Light-Emitting Devices*,
DOI 10.1007/978-94-007-1608-7_4, © Springer Science+Business Media B.V. 2011

to the criteria that are important in our device architecture. Before device fabrication, the absorption spectrum and the morphology of the materials are studied. This is subject of Sect. 4.1. Next, transistors as well as OLEDs with field-effect-assisted electron transport comprising DFH-4T and DFHCO-4T as the electron-transporting material are investigated. Special attention is given to the achievable η_{ext} at high current densities. In Sect. 4.2 transistors and OLEDs based on DFH-4T are discussed, whereas Sect. 4.3 focuses on the performance of devices comprising DFHCO-4T. An overview of the performance of different OLEDs with field-effect-assisted electron transport fabricated using various electron-transporting materials is given in Sect. 4.4. The results reported in this chapter have been published in two different articles [251, 270].

4.1 Characterization of the Electron-Transporting Materials DFH-4T and DFHCO-4T

The first material that has been investigated is DFH-4T, which was developed for use in electron-conducting thin-film transistors [269]. The molecular structure of DFH-4T can be found in Chap. 2 (Sect. 2.1.1). The material has a LUMO of 3.3 eV [237], which is slightly lower than the LUMO of Alq_3 (2.8 eV) [235] and DCM_2 (3.1 eV) [235]. From AFM analysis of thin DFH-4T films deposited by vacuum thermal evaporation, typical Stransky-Krastanov growth is observed [271, 272]. This implies that the growth of the first few DFH-4T monolayers is two-dimensional (2D). However, beyond a certain thickness threshold the film starts to roughen strongly and the growth becomes three-dimensional (3D). In the case of DFH-4T films, this 3D growth manifests itself in the formation of thick elongated needles. Figure 4.1(a) shows an AFM topography image of a 50-nm thick (as measured by a quartz crystal monitor) DFH-4T film grown on PTAA. At the bottom of the film the presence of terraces is clearly visible. Their step height is about 6 nm, corresponding to the length of two molecules [272]. On top of this 2D layer, 3D needles (up to 200 nm high) are observed. Figure 4.1(b) shows the AFM gradient image of the same DFH-4T layer. This is quite equivalent to the topography image shown in Fig. 4.1(a), however, the 2D areas and 3D needles can be recognized more clearly.

We found that the size of the 3D needles strongly depends on the deposition rate. Increasing the deposition flux reduced their height from ~300 nm to ~100 nm. The density of the 3D features could be controlled by the substrate temperature. The nature of the substrate on the other hand, apparently did not strongly influence the morphology of 50-nm thick films, which is in agreement with previous results [272]. In Fig. 4.2, three-dimensional AFM images of 50-nm thick DFH-4T films grown on top of PTAA at different deposition rates and different substrate temperatures are shown. When DFH-4T is grown at 0.2 Å/s very tall (between 300 nm and 400 nm) needle structures on top of a nice 2D layer can be observed. Increasing the substrate temperature does not really affect the height of the needles but lowers their density.

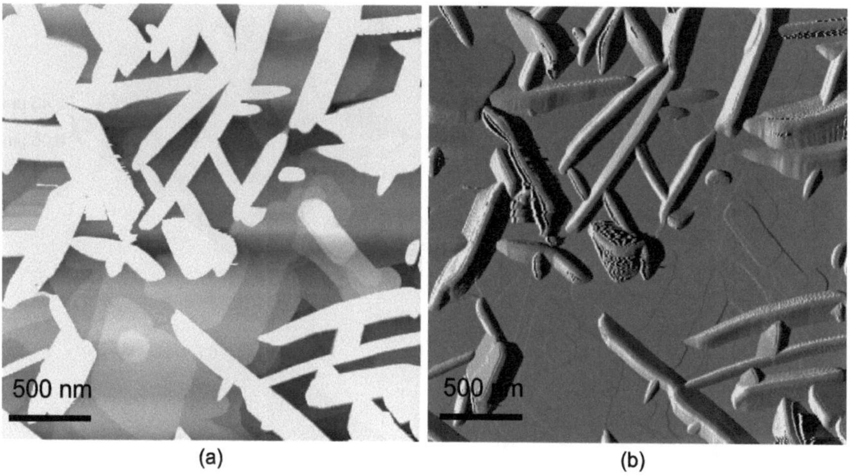

Fig. 4.1 (**a**) 2.5 μm × 2.5 μm AFM topography image of a 50-nm thick DFH-4T film grown on PTAA. A very flat 2D layer can be identified at the background. On top of this layer, large 3D needles are observed. (**b**) 2.5 μm × 2.5 μm AFM gradient image of the same layer. The 2D and 3D areas can be clearly observed. (© Wiley-VCH Verlag GmbH & Co. KGaA. Reproduced with permission)

Figure 4.2(c) and (d) show a layer of DFH-4T grown at 3 Å/s. In this case the height and the lateral size of the needles appear to be smaller. Their height is typically about 100 to 200 nm. This indicates that the transition from layered- to needle-growth can be delayed by increasing the deposition flux as well as by increasing the substrate temperature. Because a reduced height of the 3D features improves the quality and the homogeneity of the interface between the metal contact and the organic semiconductor, we typically used a deposition rate of >3 Å/s for device fabrication.

An important requirement to obtain high performance light-emitting devices is that the emitted photons are not re-absorbed by one of the other materials that is used to fabricate the device. DFH-4T is a wide-bandgap organic semiconductor, with an optical gap of 2.88 eV [273]. Thanks to its wide optical bandgap, negligible absorption is expected at the emission wavelength of DCM_2, making DFH-4T a promising candidate for use in OLEDs with field-effect electron transport. This is confirmed in Fig. 4.3, where we compare the measured normalized absorption spectrum of DFH-4T with the measured photoluminescence spectrum of Alq_3:DCM_2.

Another electron-transporting material, which has a wide band gap and therefore exhibits low absorption in the red spectral region is DFHCO-4T. The normalized absorption spectrum of DFHCO-4T, which is clearly blue-shifted with respect to the emission of DCM_2, is also shown in Fig. 4.3. In contrast to DFH-4T, its LUMO (3.9 eV) [237] is much lower than the LUMO of PTCDI-$C_{13}H_{27}$. The molecular structure of DFHCO-4T can be found in Chap. 2 (Sect. 2.1.1).

Similar to DFH-4T, AFM analysis of DFHCO-4T also reveals Stransky-Krastanov growth mode with the first few monolayers grown in a 2D arrangement

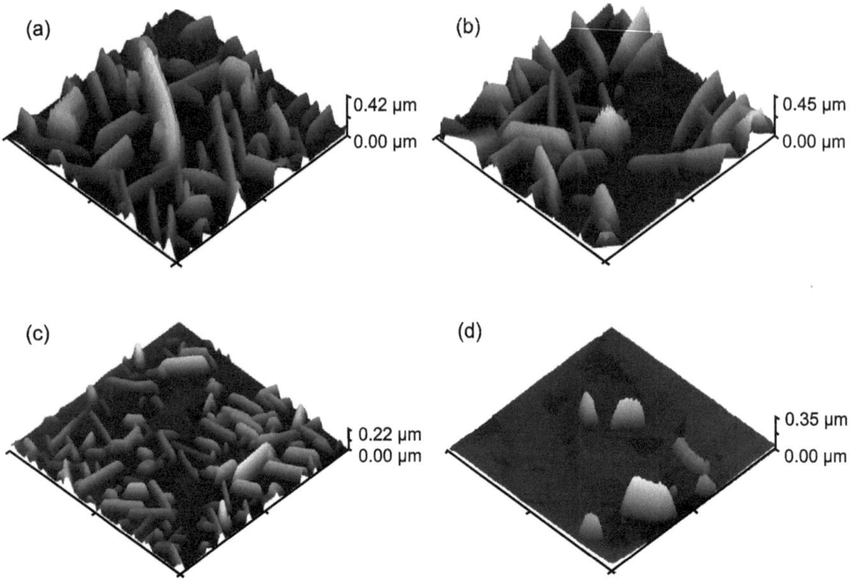

Fig. 4.2 2.5 µm × 2.5 µm three-dimensional AFM images of 50-nm thick DFH-4T films grown on top of PTAA at different deposition rates and different substrate temperatures: (**a**) flux = 0.2 Å/s and T_{sub} = 20°C, (**b**) flux = 0.2 Å/s and T_{sub} = 60°C, (**c**) flux = 3 Å/s and T_{sub} = 20°C, and (**d**) flux = 3 Å/s and T_{sub} = 60°C. (© Wiley-VCH Verlag GmbH & Co. KGaA. Reproduced with permission)

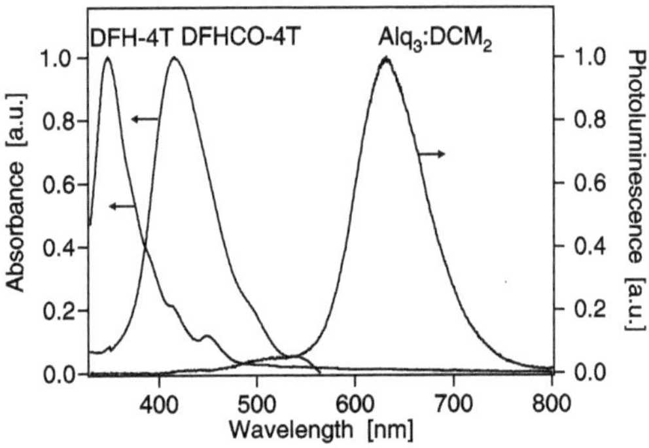

Fig. 4.3 Normalized absorption spectra of thin films of DFH-4T and DFHCO-4T and normalized photoluminescence spectrum of Alq_3:DCM_2. The low absorption in the red spectral region makes DFH-4T and DFHCO-4T particularly interesting for use in OLEDs with field-effect electron transport

followed by a 3D lattice. Also here, the transition from 2D to 3D growth can be delayed by increasing the deposition flux as well as by increasing the substrate temperature. Figure 4.4 illustrates how the deposition conditions affect the morphology of the DFHCO-4T films. Figure 4.4(a) shows an AFM image of a 50-nm thick DFHCO-4T film grown on top of PαMS at a low deposition flux (0.2 Å/s) while the substrate was kept at room temperature. Under these conditions, a very rough morphology dominated by high peaks typical for 3D growth is observed. On the other hand, at high flux and high substrate temperature the film morphology becomes much smoother, characterized by large 2D grains with only a few 3D fea-

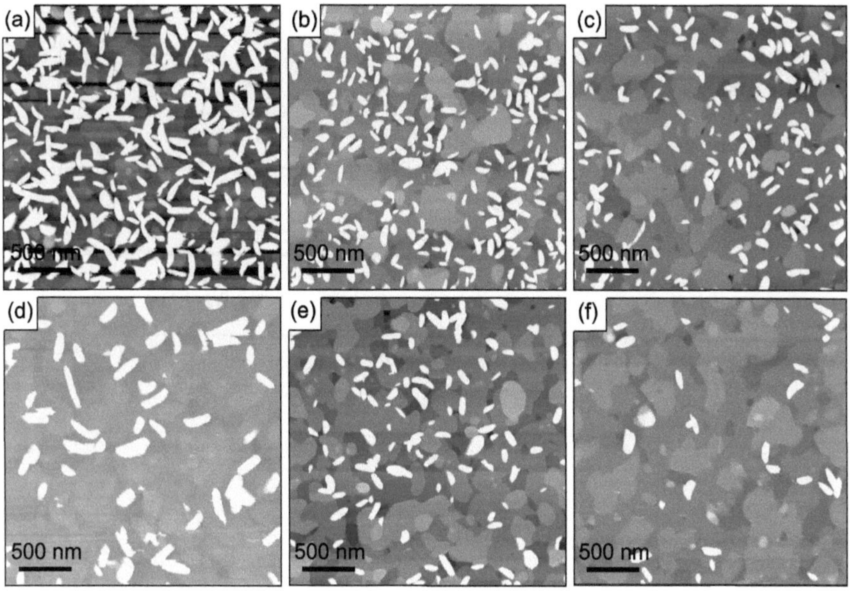

Fig. 4.4 2.5 μm × 2.5 μm AFM topography images of 50-nm thick DFHCO-4T films grown on top of PαMS at different deposition rates and different substrate temperatures: (**a**) flux = 0.2 Å/s and $T_{sub} = 20$°C, (**b**) flux = 3 Å/s and $T_{sub} = 20$°C, (**c**) flux = 4.5 Å/s and $T_{sub} = 20$°C, (**d**) flux = 0.2 Å/s and $T_{sub} = 70$°C, (**e**) flux = 3 Å/s and $T_{sub} = 70$°C, and (**f**) flux = 4.5 Å/s and $T_{sub} = 70$°C

tures (deposition rate of 4.5 Å/s and substrate temperature of 70°C, Fig. 4.4(f)). Consequently, these latter growth conditions are used for device fabrication.

4.2 Devices Based on DFH-4T

4.2.1 Transistors Based on DFH-4T

To characterize the field-effect electron transport properties of DFH-4T, we have fabricated transistors using DFH-4T as the organic semiconductor on a 100-nm thick SiO_2 dielectric. LiF/Al top contacts were used as source and drain electrodes. Prior to the deposition of DFH-4T, PTAA was spincoated. We have demonstrated earlier on that the hole-transporting wide-bandgap organic semiconductor PTAA provides a good growth surface for electron-conducting organic semiconductors, and that the hetero-interface is appropriate for electron transport. The output and transfer characteristics of these DFH-4T transistors are shown in Fig. 4.5. The effective saturation field-effect mobility for electrons at the heterojunction interface with PTAA is 0.15 cm^2/Vs. This is similar to the electron mobility we obtained for PTCDI-$C_{13}H_{27}$. We correlate this high electron field-effect mobility to the good 2D growth of the first monolayers of DFH-4T when grown on top of PTAA [274, 275].

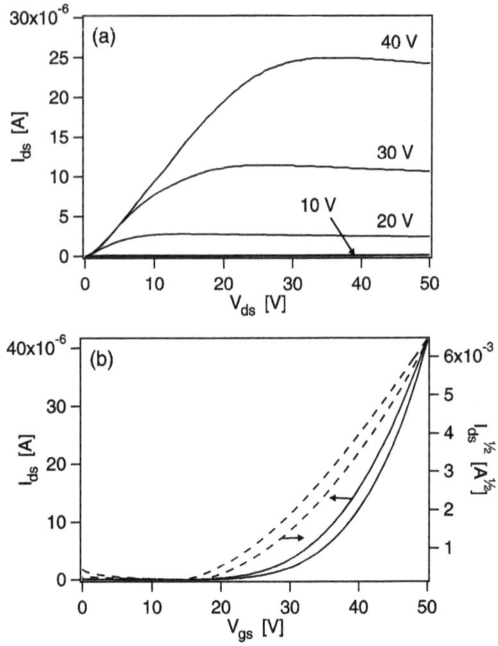

Fig. 4.5 Electrical characteristics of a DFH-4T top contact transistor ($W/L = 2000/50$): (**a**) Output characteristics, and (**b**) transfer characteristics. The saturation mobility for electrons and the threshold voltage are, respectively, 0.15 cm^2/Vs and ~24 V

To evaluate the electron transport properties of DFH-4T in OLEDs with field-effect electron transport, we also fabricated and characterized transistors using an additional Alq_3 layer in between the PTAA and the DFH-4T. Upon inserting Alq_3, the electron mobility of DFH-4T slightly dropped to a value of 0.03 cm^2/Vs.

4.2.2 DFH-4T in OLEDs with Field-Effect Electron Transport

OLEDs with field-effect electron transport comprising PTAA as the organic hole-transporting material, Alq_3:DCM_2 as the organic light-emitting layer and DFH-4T as the organic electron-transporting material have been fabricated. ITO is used as the hole-injecting electrode in the device and the cathode is formed by a thin layer of lithium fluoride (LiF) followed by deposition of aluminum (Al). Figure 4.6 displays the experimentally measured current-voltage characteristics of a device that has a width of 1 mm and where the metallic cathode was displaced by 7.2 μm with respect to the insulator edge. The measured maximum η_{ext} of the device is 0.25% (Fig. 4.13), which is more than 10 times larger than the value we obtained for devices employing 50 nm PTCDI-$C_{13}H_{27}$ as the electron-transporting material (Sect. 3.3). Clearly, this improved η_{ext}, which is approaching the efficiency of reference OLEDs with Alq_3:DCM_2 active layers [254], can be explained by the reduced absorption of the emitted photons by the electron-transporting material.

Under forward bias, the current increases with increasing voltage. However, in contrast to PTCDI-$C_{13}H_{27}$-based devices, the current appears to flatten out at volt-

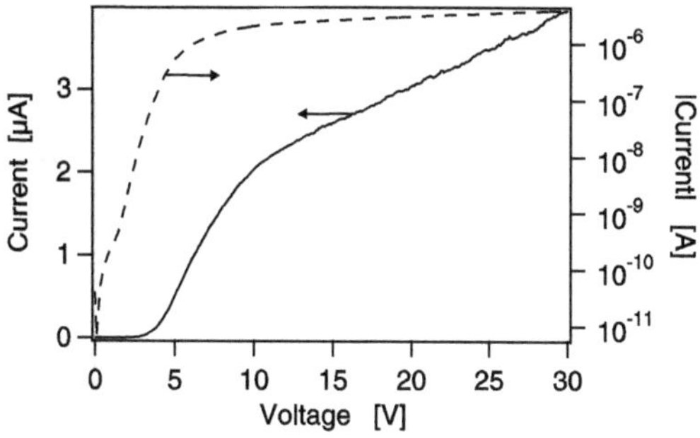

Fig. 4.6 Current-voltage characteristics of a 1 mm wide OLED with field-effect-assisted electron transport comprising the wide bandgap material DFH-4T as the electron-transporting material. The distance between the metallic top contact and the insulator edge is 7.2 μm. The *dashed line* represents the current on a logarithmic scale, whereas the *full line* is the current plotted on a linear scale. (© Wiley-VCH Verlag GmbH & Co. KGaA. Reproduced with permission)

ages exceeding 10 V. This limits the maximum current density in the device. In order to calculate the maximum electron current density, we have to consider the thickness of the accumulation layer. Since field-effect transport typically occurs in the first few monolayers of the semiconductor film [45, 276–278], we assume that the electron current flow is confined to a 6 nm thick electron accumulation layer, which corresponds to approximately two DFH-4T monolayers [272]. Taking into account this accumulation layer thickness, the maximum electron current density at 30 V is estimated to be $\sim$60 A/cm^2. This is much less than what we would expect when the current increases more or less quadratically with the applied voltage as it is the case in PTCDI-$C_{13}H_{27}$-based devices. In order to understand what limits the current increase, several control devices were measured. We fabricated devices using the organic stacks PTAA/DFH-4T and PTAA/Alq_3/DFH-4T. In both cases no flattening out of the current at higher voltages was observed. From these measurements we infer that the slower increase of the current with voltage for voltages above 10 V in Fig. 4.6 is caused by the field distribution in the device at these high voltages combined with the presence of DCM_2.

Figures 4.7(a) and (b) show the band diagrams for the device in forward bias, i.e. for a positive anode-to-cathode bias, respectively, at low and high voltages. Under these conditions electrons are injected from the cathode into DFH-4T. At low voltage bias (Fig. 4.7(a)), both the PTAA and the light-emitting layer Alq_3:DCM_2 on top of the SiO_2 are depleted of holes. The LUMO offset at the interface between Alq_3:DCM_2 and DFH-4T prohibits electron injection from DFH-4T into Alq_3:DCM_2. As a result, the depleted PTAA and Alq_3:DCM_2 layers on top of the SiO_2 insulator behave as a dielectric layer in series with the SiO_2 insulator, and an electron accumulation layer is formed in the DFH-4T at the interface with Alq_3:DCM_2. However, at higher voltage bias (Fig. 4.7(b)), the energy bands are much more bent. Since the LUMO of DFH-4T (3.3 eV) [237] is slightly higher than that of PTCDI-$C_{13}H_{27}$ (3.4 eV) [236], we assume that at high voltage bias under the influence of the electric field some electrons can overcome the energy barrier

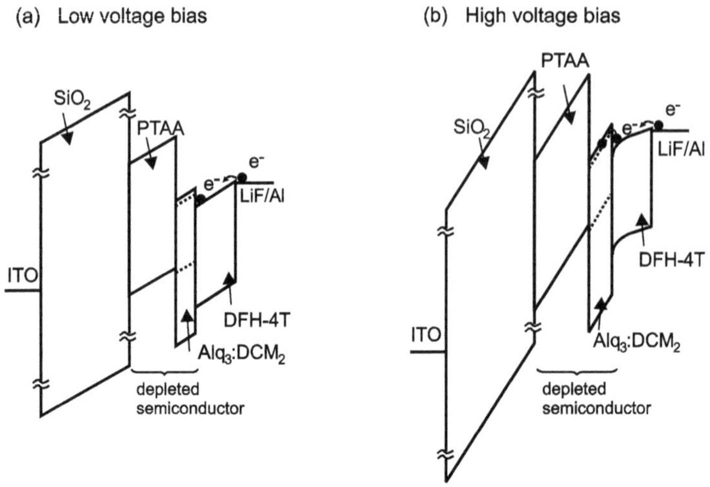

Fig. 4.7 Band diagrams for the device under forward bias conditions: (**a**) at low voltage bias, and (**b**) at high voltage bias. (© Wiley-VCH Verlag GmbH & Co. KGaA. Reproduced with permission)

and are trapped on DCM_2. This fixed charge yields, with the same field over the dielectric, less mobile charge in the channel resulting in a lower current.

This hypothesis was confirmed by measuring the characteristics of a device where a thin (10 nm) undoped layer of Alq_3 was inserted in between the Alq_3:DCM_2 light-emitting layer and the DFH-4T electron-transporting layer. In this device no flattening out of the current at high voltages was observed (Fig. 4.8) and as a result high electron current densities of 750 A/cm^2 were obtained at 30 V. η_{ext} remains unaffected upon insertion of this thin undoped Alq_3 layer (Fig. 4.13). An additional argument supporting our hypothesis is the fact that similar flattening out of the current is observed in PTCDI-$C_{13}H_{27}$-based devices when biases exceeding 40 V are applied. Since the LUMO of PTCDI-$C_{13}H_{27}$ is lower than the one of DFH-4T it is obvious that higher voltages are needed to overcome the energy barrier and to trap electrons.

4.3 Devices Based on DFHCO-4T

4.3.1 High Performance DFHCO-4T Transistors

An electron field-effect mobility of 1.7 cm^2/Vs has been reported by Yoon *et al.* for the organic semiconducting material DFHCO-4T [272]. In these devices a top-contact geometry with Au source and drain contacts was used. To check whether LiF/Al could also be used as top contact material we have fabricated DFHCO-4T transistors with LiF/Al top contacts (0.8 nm LiF followed by 100 nm Al) and compared their performance with DFHCO-4T transistors having 100 nm Au as the top

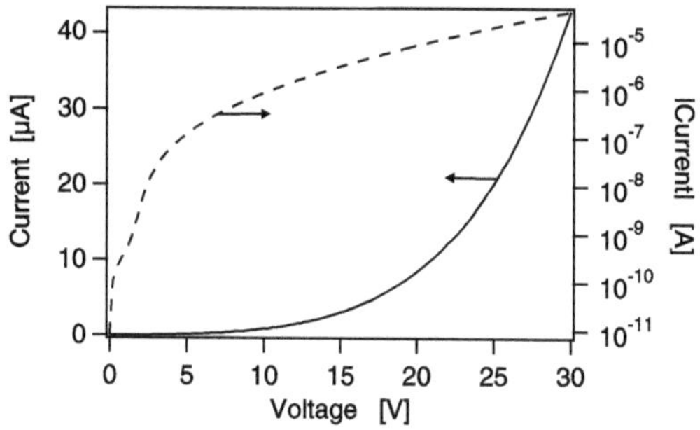

Fig. 4.8 Electrical characteristics of an OLED with field-effect-assisted electron transport comprising PTAA/Alq_3:DCM_2/Alq_3/DFH-4T as the organic layer stack. The *dashed line* represents the current on a logarithmic scale, whereas the *full line* is the current plotted on a linear scale. The device width is 1 mm and the distance between the metallic top contact and the insulator edge is 9.7 µm. (© Wiley-VCH Verlag GmbH & Co. KGaA. Reproduced with permission)

metal contact. Thin DFHCO-4T films were deposited by thermal vacuum evaporation of the material using the optimized growth conditions (Sect. 4.1). Prior to the deposition of DFHCO-4T, a 5-nm thick PαMS layer was spincoated on top of the Si/SiO_2 substrate to provide a high-quality, electron-trap free surface allowing excellent electron transport [259].

Figure 4.9(a) shows the output characteristics of a transistor with 130 µm channel length and LiF/Al top contacts. For this device, an apparent field-effect electron mobility of 0.03 cm^2/Vs was calculated. This value is more than one order of magnitude lower than the electron mobility of DFHCO-4T previously reported using Au top contacts [272]. In addition, the OTFTs show strongly non-ideal characteristics, as apparent from a superlinear dependence of the current on the drain voltage at low bias, the saturation of the output current at a more or less fixed drain-to-source bias, and a decrease of the current at higher drain voltages.

On the other hand, transistors fabricated under the same DFHCO-4T growth conditions but with Au instead of LiF/Al source-drain top contacts exhibit much better-behaved output characteristics with much higher drain currents (Fig. 4.9(b)). For Au top-contact transistors with 130 µm channel length, a mobility of 4.6 cm^2/Vs is measured. Such a remarkable mobility achieved with Au contacts is of high technological relevance because for use in complementary logic it is preferable to use a single type of source and drain metal for both the p-type and the n-type OTFTs. The reproducibility of the results is quite good. The average apparent mobility, calculated from transistors with channel lengths between 50 and 200 µm is 3.5 cm^2/Vs. These field-effect mobilities are higher compared to previously reported results [272] and we attribute this to the different growth conditions used to deposit DFHCO-4T. The fact that electrons can be efficiently injected from Au into DFHCO-4T despite an injection barrier of $\sim$1 eV is not fully understood yet. The deep LUMO (3.96 eV) [237] of DFHCO-4T and the fact that in a top-contact geometry there is a high gate-field that supports the injection of charges in a large source-gate overlap area [67] might be a possible explanation for this experimental observation.

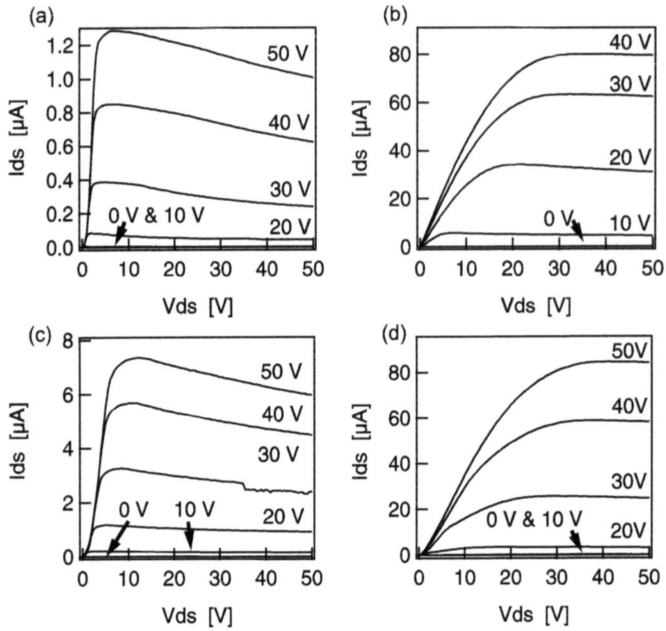

Fig. 4.9 Output characteristics of DFHCO-4T transistors ($W/L = 2000/130$) using (**a**) LiF/Al, (**b**) Au, (**c**) Yb, and (**d**) Ag top contacts. (© American Institute of Physics. Reproduced with permission)

The low apparent field-effect mobility (i.e., low output currents) and the non-ideal characteristics of DFHCO-4T transistors with LiF/Al top contacts compared to Au top contacts point to the presence of a current-limiting contact problem in the former. A plausible cause of this is the occurrence of a redox reaction between the n-type semiconductor and the Al metal contact. In fact, it is well known that ketone groups—present in the DFHCO-4T chemical structure—are easily reduced into their radical anions according to the reaction scheme shown in Fig. 4.10(a) [279]. The electron transfer reaction is driven by the oxidation of a reactive metal, as for example Mg, Al, Fe and Zn (depicted in Fig. 4.10(b) for Al) [280, 281] and the reduction of DFHCO-4T. The standard free energy ($\Delta G°$, Gibbs free energy) of the overall redox reaction depicted in Fig. 4.10(c) can be calculated by Eq. 4.1 [282], where n is the overall number of electrons exchanged between the oxidizing and reducing agents for the balanced redox equation, F the Faraday constant (96485 J/(V mol)), and $\Delta E°$ the standard cell potential (the difference in standard electrode potentials (SEPs) of both electrochemical couples). An overview of the SEPs of various oxidizing and reducing agents measured in solution is given in Fig. 4.11.

$$\Delta G° = -nF\Delta E°. \tag{4.1}$$

For the Al/Al^{3+} couple the SEP in aqueous solution is -1.662 V vs the normal hydrogen electrode (NHE) [283], which is about -1.90 V vs the reference saturated calomel electrode (SCE) [284]. Published value for the first reduction potential of

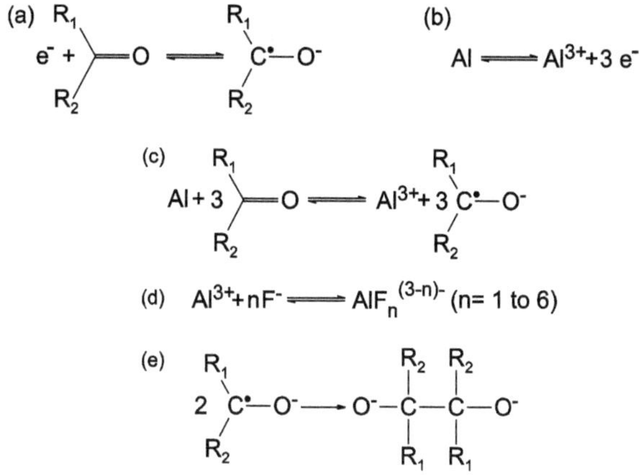

Fig. 4.10 (**a**) Reduction of a ketone, (**b**) oxidation of Al, (**c**) redox reaction between a ketone and Al, (**d**) complexation of Al^{3+} by F^-, and (**e**) dimerization of a ketone radical anion. (© American Institute of Physics. Reproduced with permission)

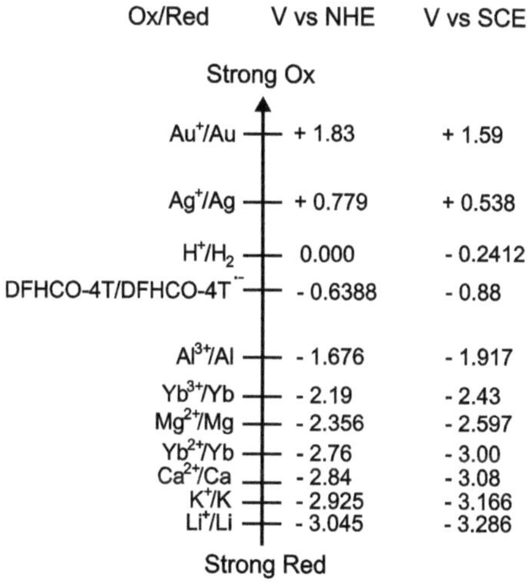

Fig. 4.11 Overview of the standard electrode potential of various electrochemical couples measured in solution

DFHCO-4T in tetrahydrofuran is -0.88 V vs the SCE [237]. Inserting this potential difference into Eq. 4.1, with $n = 3$ (Fig. 4.10(c)), gives $\Delta G^\circ = -296$ kJ/mol. This indicates that the reaction between Al and DFHCO-4T is highly exergonic and thus thermodynamically spontaneous.

Similar reductions of organic layers by evaporation of strongly reducing metals have been reported in literature. The reaction of Al evaporated on films of Alq_3 [285–288], $F_{16}CuPc$[1] [289] and PTCDA[2] [290] for example were experimentally proven. The reduction of DFHCO-4T by evaporated Al (Fig. 4.10(c)) in our case is furthermore favored by additional side reactions. One of these reactions is the complexation of the Al^{3+} cation by fluoride ions (F^-) originating from an ultra-thin (0.8 nm) LiF layer on top of the DFHCO-4T. In fact, F^- anions are

[1] Copper hexadecafluorophthalocyanine.

[2] 3,4,9,10 perylenetetracarboxylic dianhydride.

well known to form stable complexes ($AlF_n^{(3-n)+}$ with $n = 1$–6, Fig. 4.10(d)) with Al^{3+} cations [291]. Furthermore, ketone radical-anions readily dimerize affording the corresponding acyloinic species (Fig. 4.10(e)) [280]. Such additional processes will enforce the DFHCO-4T/Al redox reaction (Fig. 4.10(c)) towards the right side according to Le Chatelier's principle.

The products of the reactions displayed in Fig. 4.10(c–e), an ionic salt and possibly dimerized ketone species, will form a thin interfacial layer between the unreacted DFHCO-4T and Al layers, hindering electron injection. In the case of Au this kind of reaction is impossible: the SEPs of the couples involving oxidation of Au are so high (+1.692 V vs NHE for Au/Au^+ and +1.492 V vs NHE for Au/Au^{3+}) [283] that a reaction with DFHCO-4T would be highly endergonic ($\Delta G^\circ = +225$ kJ/mol in the case of Au/Au^+ and +619 kJ/mol for Au/Au^{3+}).

To find additional proof supporting the chemical reaction between Al and DFHCO-4T, we fabricated transistors using DFHCO-4T as the organic semiconductor but with two other top-contact metals: Ytterbium (Yb) and silver (Ag). The standard free energies for possible reactions between DFHCO-4T and Yb indicate that this kind of reaction is highly exergonic and thus spontaneous, independently if the Yb/Yb^{2+} couple (SEP: −2.76 V vs NHE [283], corresponding to $\Delta G^\circ = -409$ kJ/mol) or Yb/Yb^{3+} couple (SEP: −2.19 V vs NHE [283], corresponding to $\Delta G^\circ = -449$ kJ/mol) is considered. On the other hand, the corresponding value for the reaction of DFHCO-4T with Ag (SEP: +0.7796 V vs NHE [283] for Ag/Ag^+) is +139 kJ/mol, indicating that this reaction is impossible.

The output characteristics of Yb and Ag top-contact transistors with 130 μm channel length are shown in Fig. 4.9(c) and (d). Transistors with Yb top contacts show similar output characteristics and low apparent field-effect mobility (0.06 cm^2/Vs) as LiF/Al transistors, whereas Ag top-contact transistors yield much higher currents and much better-behaved output characteristics. When Ag is used as top metal contact an apparent mobility of 1.7 cm^2/Vs was measured. These results are in agreement with our argumentation and give further evidence that a redox reaction occurs at the interface between DFHCO-4T and easily oxidizable metals, similar to Hirose *et al.*'s report that PTCDA reacts with Al, but not with Au and Ag [290].

4.3.2 DFHCO-4T in OLEDs with Field-Effect Electron Transport

The high electron field-effect mobility that can be obtained in DFHCO-4T combined with the low LUMO and the low absorption in the red spectral region makes DFHCO-4T particularly interesting for use as the electron-transporting material in OLEDs with field-effect electron transport. Devices comprising PTAA as the hole-transporting layer, Alq_3:DCM_2 as the light-emitting layer and DFHCO-4T as the electron-transporting layer have been fabricated. Because DFHCO-4T only functions properly with contacts that have a low chemical reactivity such as Au and Ag, we used the inert contact metal Au to inject electrons in the device.

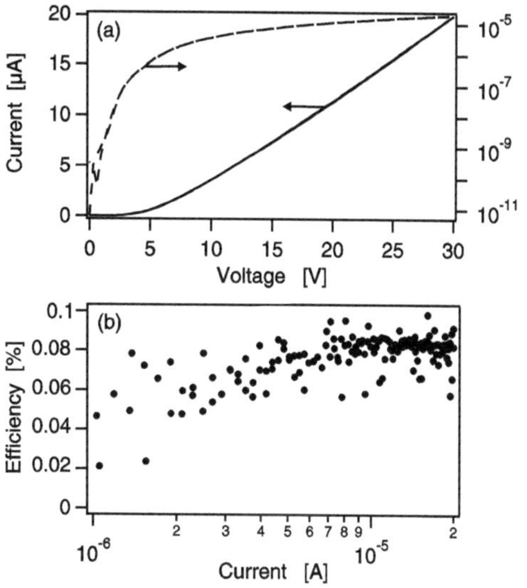

Fig. 4.12 Measured characteristics of an OLED with field-effect-assisted electron transport using the wide band gap material DFHCO-4T as the electron-transporting material. The device has a width of 1 mm and a distance between the top electrode and the insulator edge of 6.7 µm. (**a**) Current-voltage characteristics, and (**b**) η_{ext} as a function of the current

The current-voltage characteristics, the light intensity and the electroluminescence spectrum of the devices were measured. The results are summarized in Fig. 4.12. An η_{ext} of 0.08% was calculated, which is lower than the value obtained in devices using DFH-4T, but still significantly higher than the one of PTCDI-$C_{13}H_{27}$-based devices. Comparable to previous results, there is no significant reduction of η_{ext} up to the maximum current.

4.4 Comparison

An overview of the performance of different OLEDs with field-effect-assisted electron transport fabricated using various electron-transporting materials is given in Fig. 4.13. In this figure the external quantum efficiencies η_{ext} as a function of the current for devices comprising PTCDI-$C_{13}H_{27}$, DFH-4T or DFHCO-4T as the electron-transporting layer are compared. All devices presented in Fig. 4.13 have the same distance to the insulator edge (~7 µm) and were biased up to 30 V. By replacing the 50-nm thick PTCDI-$C_{13}H_{27}$ layer by DFH-4T a tenfold increase in the external quantum efficiency could be obtained. This improvement is mainly attributed to the reduced light re-absorption in the red spectral region where DCM_2 is emitting. The insertion of a thin undoped Alq_3 layer in between the DFH-4T and the active Alq_3:DCM_2 light-emitting layer does not affect this efficiency. Moreover, the presence of this undoped Alq_3 layer makes it possible to achieve high current densities in the device.

Within the error margin of about 20% we can say that PTCDI-$C_{13}H_{27}$-based devices yield a three times higher current than DFH-4T-based devices having the same displacement with respect to the insulator edge. We attribute this to the difference in mobility between PTCDI-$C_{13}H_{27}$ (0.08 cm^2/Vs) and DFH-4T (0.03 cm^2/Vs)

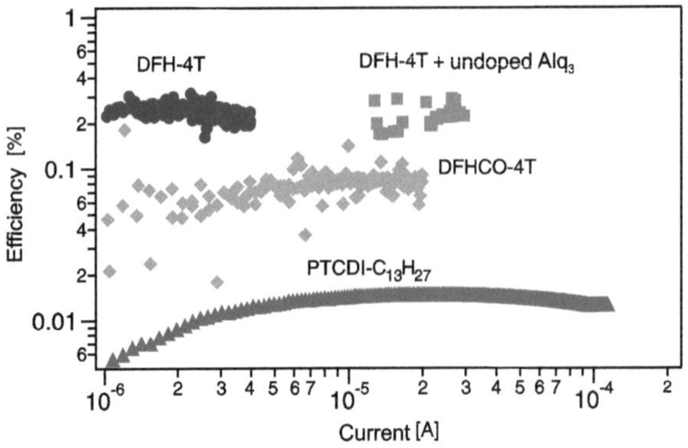

Fig. 4.13 η_{ext} as function of the current for different devices using, respectively, 50 nm of PTCDI-$C_{13}H_{27}$ (*triangles*), 50 nm of DFH-4T (*circles and squares*) or 50 nm of DFHCO-4T (*diamonds*) as the electron-transporting layer. The rectangular markers represent the η_{ext} of a device where a thin undoped Alq_3 layer is inserted in between the DFH-4T and the active light-emitting layer. All four devices have a width of 1 mm and a distance to the insulator edge of ~7 μm

when grown on top of Alq_3. On the other hand, contrary to the higher field-effect electron mobility measured in DFHCO-4T-based devices (0.8 cm^2/Vs on top of Alq_3:DCM_2), a lower current is measured. This is probably related to the very low LUMO of DFHCO-4T (3.9 eV compared to 3.3 eV for DFH-4T and 3.4 eV for PTCDI-$C_{13}H_{27}$). This low LUMO creates a large injection barrier for electrons, limiting the current through the device and prohibiting efficient recombination of holes and electrons.

4.5 Summary and Conclusions

In this chapter, two high field-effect electron-mobility organic semiconductors, DFH-4T and DFHCO-4T, were studied. Both materials have a wide band gap and show low absorption in the red spectral region. The growth of DFH-4T and DFHCO-4T was optimized and transistors using these materials as the electron-conducting semiconductor were fabricated. In addition, we compared the performance of DFHCO-4T transistors with different top-contact metals. While most n-type organic thin-film transistors show good performance when using LiF/Al as top-contact material [255, 292], we found that DFHCO-4T top-contact transistors only function properly with contacts that have a low chemical reactivity such as Au and Ag. Mobilities as high as 4.6 cm^2/Vs were demonstrated for DFHCO-4T transistors with Au top contacts, whereas the mobility of similar transistors fabricated with easily oxidizable top-contact metals, such as LiF/Al and Yb, was much lower. The reduced performance of LiF/Al and Yb top-contact transistors was attributed to an electron-transfer reaction occurring at the metal/DFHCO-4T interface. Since Au and Ag—in contrast to Al and Yb—do not react with the semiconducting material, the use of these inert metals is beneficial to achieve high-performance DFHCO-4T-based devices.

The use of DFH-4T and DFHCO-4T in OLEDs with field-effect-assisted electron transport was also investigated. Due to the reduced light re-absorption in the red spectral region where DCM_2 is emitting, higher external quantum efficiencies could be achieved compared to PTCDI-$C_{13}H_{27}$-based devices. For devices using DFH-4T as the electron-transporting material, an η_{ext} of 0.25% was demonstrated. In addition, we showed that by optimizing the growth of DFH-4T at organic hetero-interfaces, and by properly engineering the material stack of the Alq_3:DCM_2 host-guest light-emitting system to avoid carrier trapping at the guest molecules, high current densities up to 750 A/cm^2 could be achieved in the electron channel of these DFH-4T-based OLED devices. OLEDs with field-effect-assisted electron transport comprising DFHCO-4T as the electron-transporting material exhibited an η_{ext} of 0.08%. For these devices an electron current density of about 300 A/cm^2 was measured.

Chapter 5
Control of the Triplet Concentration in Organic Light-Emitting Devices

The vivid interest in organic semiconductor materials is stimulated by the technological aspects of applications that employ these new materials in OLEDs [23, 27], thin-film field-effect transistors [93–95], solar cells [82] and biological sensors. Some products, such as emissive OLED displays, [28] have gained rapid acceptance in portable displays because of their bright emission and low power consumption. The remarkable technological breakthroughs in the development of highly efficient OLEDs and the demonstration of optically-pumped lasing [56, 90] using the same class of organic semiconductors have focused an increasing interest in the prospects for making electrically-pumped organic solid-state lasers [85, 86]. However, as discussed in Chap. 1 (Sect. 1.3.4), the combination of several issues makes electrical excitation a particularly tough problem for organic semiconductor lasers. One of the main problems associated with electrically pumped organic semiconductor lasers and other high-brightness organic light sources is the inevitable population and accumulation of triplet excitations, which result in excessive triplet-state losses, such as triplet-triplet absorption [223, 228] and singlet-triplet annihilation [50, 212, 215].

In this chapter, two different approaches to reduce the triplet concentration in organic light-emitting devices are discussed. A first method to control triplet accumulation is by operating the device under pulsed excitation [228]. This is the subject of Sect. 5.1. The behavior of OLEDs with field-effect-assisted electron transport is analyzed under pulsed excitation and their performance is compared to conventional OLEDs using the same materials. If the transit time of carriers through transport layers contributes to the determination of the minimum pulse duration applicable to an OLED, it is expected that devices employing an enhanced field-effect mobility may allow shorter pulse durations compared to standard OLEDs. Section 5.2 then focuses on an alternative way to control the triplet concentration. Polymer films doped with appropriate triplet scavengers are studied by time-resolved photoluminescence measurements. Vertical as well as nonvertical triplet scavengers are investigated and their use for triplet reduction in organic light-emitting devices is argued. The content of this chapter has been published in two different articles [293, 294].

S. Schols, *Device Architecture and Materials for Organic Light-Emitting Devices*,
DOI 10.1007/978-94-007-1608-7_5, © Springer Science+Business Media B.V. 2011

5.1 Pulsed Excitation of OLEDs with Field-Effect Electron Transport

In literature, pulsed operation is suggested as being one of the approaches to reduce triplet-state losses in organic light-emitting devices [225, 226]. Pulsed excitation makes it possible to separate singlet excitons from triplet excitons in the time domain, allowing a tremendous reduction of the triplet concentration and associated losses in an electrically driven organic light-emitting device [228]. In addition, pulsed operation avoids heat and stress in the device [227] and it allows higher current densities compared to continuous wave (CW) operation [209]. In this section, the behavior of OLEDs with field-effect-assisted electron transport under pulsed excitation is investigated.

5.1.1 Pulse-Width Dependence of Organic Light-Emitting Devices

The devices used in this study comprised 50 nm PTAA as the hole-transporting layer, 20 nm Alq_3:DCM_2 as the light-emitting layer and 50 nm PTCDI-$C_{13}H_{27}$ as the electron-transporting layer. Standard OLEDs as well as OLEDs with field-effect-assisted electron transport were fabricated. The devices were characterized in an inert N_2 atmosphere immediately after evaporation of the metallic top-contact. An Agilent 8114A pulse generator (rise and fall time = 8 ns) was used to apply rectangular pulses to the devices. The duty cycle was 10%, and the pulse length was varied between 100 ns and 100 μs. All coax cables were 50 Ω terminated to avoid reflections. A calibrated integrated sphere (SphereOptics Hoffman GmbH) coupled to an Agilent 4156C parameter analyzer was used for light intensity measurements.

Figure 5.1 shows the measured peak light intensity as a function of the pulse width for a conventional OLED and an OLED with field-effect-assisted electron transport biased at, respectively, 7 V and 20 V. Both devices were fabricated during the same evaporation runs and use the same organic layer stack. From Fig. 5.1 the minimum pulse width that can be applied without affecting the light output can be determined. Both devices allow pulse widths down to ~10 μs without significant reduction of the light intensity. The shape of the curves, however, differs strongly. The OLED with field-effect electron transport exhibit a significantly weaker dependence on the pulse duration compared to the conventional OLED.

Note that the minimum pulse width that can be applied without significant reduction of the light intensity is much shorter than the typical triplet lifetime of fluorescent materials (milliseconds to seconds [230]), indicating that pulsed excitation is useful to reduce the accumulation of triplets in these devices. Indeed, when the pulse repetition rate is sufficiently low and pulses shorter than the lifetime of triplet excitons are applied, the build-in of triplets can be reduced.

If the transit time of the carriers through the organic layers contributes to the determination of the minimum pulse duration, it is expected that OLEDs with field-effect-assisted electron transport, which employ an enhanced field-effect mobility,

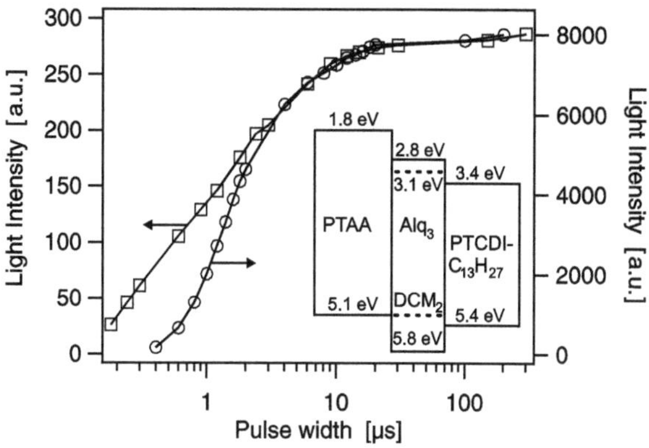

Fig. 5.1 The measured peak light intensity as a function of the pulse width for a conventional OLED (*open circles*) and an OLED with field-effect-assisted electron transport (*open squares*) biased, respectively, at 7 V and 20 V. *Inset*: energy level diagram of the organic layer stack under flat band conditions. The *dotted lines* indicate the HOMO and LUMO positions of DCM_2. (© 2010 IEEE)

could allow shorter pulse durations compared to standard OLEDs. In this way, a more effective reduction of the triplet concentration would be possible when operating the device under pulsed excitation. However, Fig. 5.1 clearly shows that there is no distinct difference between the minimum allowable pulse width without reduction of the light intensity of both device architectures.

5.1.2 Effects Influencing the Pulsed Excitation Behavior of OLEDs

Charge Transport

Different effects may influence the pulse-width dependence of organic light-emitting devices. A first possible reason for the decrease of the light output at short pulse widths could be the finite transit time of the charge carriers through the transport layers. One of the parameters controlling this process is the charge carrier mobility. If the mobility of one of the charge carriers is too low, the steady-state exciton concentration cannot be reached within the pulse duration, resulting in a decrease of the measured light intensity.

To verify whether the charge transport through the electron accumulation layer in PTCDI-$C_{13}H_{27}$ is a limiting factor, OLEDs with field-effect electron transport having different distances between the metallic contact and the insulator edge, and thus different transit times, were fabricated. The result is presented in Fig. 5.2(a): all devices showed the same pulse-width dependence.

Subsequently, it was verified whether charge transport in the hole-transporting or the light-emitting layer could be limiting, by varying the thicknesses of the PTAA and the Alq_3:DCM_2 layers. Figure 5.2(b) shows the dependence of the light intensity on the pulse width for OLEDs comprising 15 nm or 50 nm PTAA, whereas in Fig. 5.2(c) the thickness of the active light-emitting layer was changed. Figure 5.2

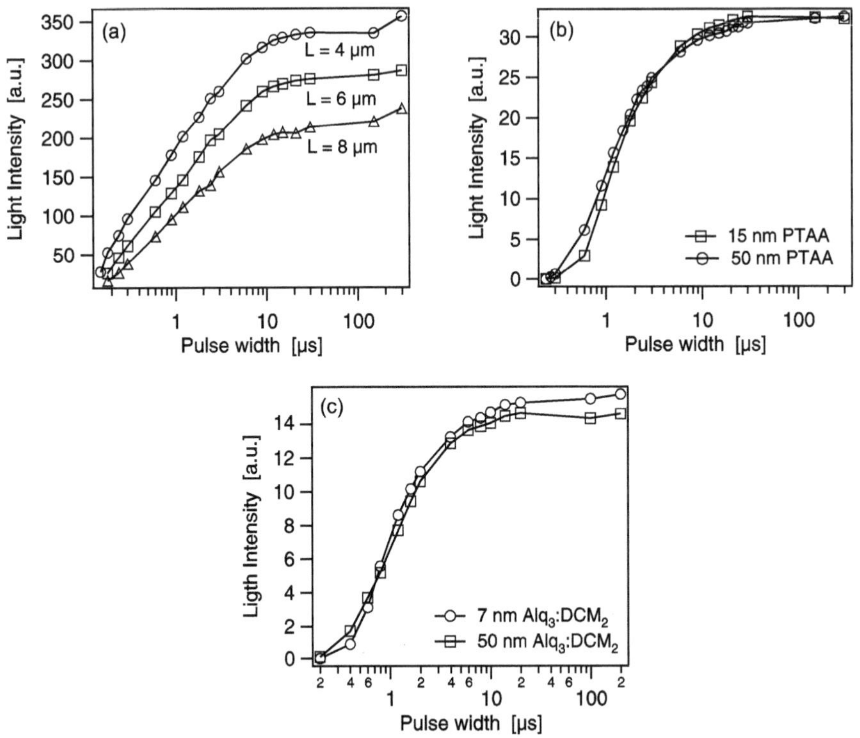

Fig. 5.2 Dependence of the peak light intensity on the pulse width for (**a**) OLEDs with field-effect-assisted electron transport with, respectively, a distance of 4 μm, 6 μm and 8 μm between the metallic contact and the insulator edge, (**b**) conventional OLEDs comprising 15 nm or 50 nm PTAA, and (**c**) conventional OLEDs having an active layer of 7 nm or 50 nm thick. The OLEDs with field-effect-assisted electron transport were biased at 20 V, whereas 7 V was applied to the conventional OLEDs. (© 2010 IEEE)

shows that the dependence of the light output on the applied pulse duration is invariant with the thickness of the layers, and hence not limited by the transit times of the charge carriers through the layers in our device structure.

Charge Accumulation

In multi-layer OLEDs, however, not only charge transport, but also charge accumulation at the hetero-interfaces between the different semiconductors should be taken into account [295]. The energy level diagram under flat band conditions, shown as inset in Fig. 5.1, clearly indicates the presence of energy level offsets at the organic/organic interfaces. Consequently, injected electrons and holes, transported through the PTCDI-$C_{13}H_{27}$ and PTAA layer, respectively, will accumulate at these interfaces before injection into the active light-emitting layer. Depending on the applied electric field, space charge will be built up faster and the charges will more

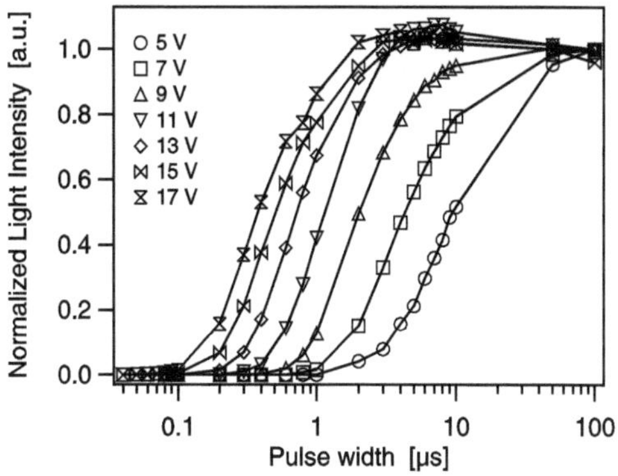

Fig. 5.3 Pulse-width dependence of a standard OLED comprising PTAA/Alq_3:DCM_2/PTCDI-$C_{13}H_{27}$ as the organic layer stack biased at different voltages. (© 2010 IEEE)

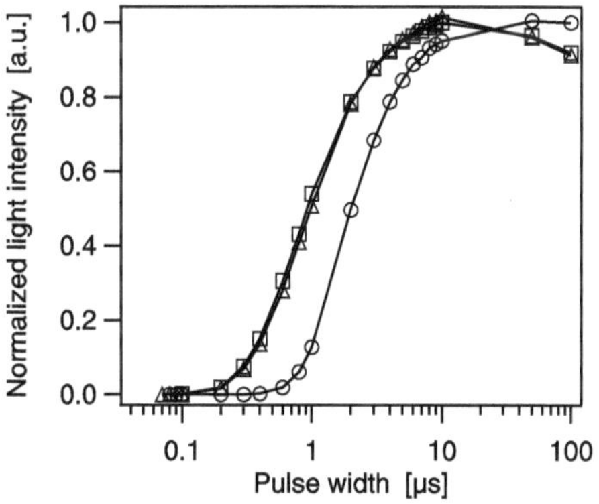

Fig. 5.4 Normalized light intensity as a function of the pulse width of OLEDs using: PTAA/Alq_3:DCM_2/PTCDI-$C_{13}H_{27}$ (*open circles*), PTAA/Alq_3:DCM_2 (*open triangles*), and PTAA/Alq_3:DCM_2/Alq_3 (*open squares*) as the organic layer stack. A bias of 9 V was applied to the devices. (© 2010 IEEE)

easily overcome the present energy barriers. The dependence of the light intensity on the pulse width for an OLED biased at different voltages is shown in Fig. 5.3. It can be clearly seen that the minimum allowable pulse width without reduction of the light intensity decreases from 10 μs to about 1 μs when higher voltages were applied. This points out that one of the organic/organic interfaces indeed limits the behavior of the device under pulsed excitation.

To verify which interface is the limiting factor two additional OLEDs were fabricated: ITO/PTAA/Alq_3:DCM_2/LiF/Al and ITO/PTAA/Alq_3:DCM_2/Alq_3/LiF/Al. In both devices the energy barrier between the electron-transporting layer and the active light-emitting layer is eliminated. The total thickness of the organic layers was kept constant in order to apply the same electric field over the devices. Consequently, the first device used 70 nm of Alq_3:DCM_2, whereas the second OLED had a 20-nm thick Alq_3:DCM_2 layer and 50 nm undoped Alq_3. Figure 5.4 compares the light output of these two OLEDs with the one of Fig. 5.3 as a function of the pulse width when a bias of 9 V is applied. A clear difference in pulsed excitation behavior is observed. The minimum pulse width that could be applied without affecting the light intensity shifted to shorter values when the energy offset at the interface between the PTCDI-$C_{13}H_{27}$ and the Alq_3:DCM_2 layer was eliminated. This indicates that it is the PTCDI-$C_{13}H_{27}$/Alq_3:DCM_2 interface that has a limiting effect on the pulsed excitation properties of our light-emitting devices. Since this interface is present in the conventional OLED as well as in the OLED with field-effect electron transport, it is obvious that the minimum allowable pulse width without reduction of the light intensity is similar for both devices (Fig. 5.1). The fact that the dependence on the pulse duration is weaker in the case of an OLED with field-effect-assisted electron transport can be understood by a spread in potential drop. Indeed, the light-emission zone of an OLED with remote cathode is approximately 2 μm wide (Sect. 3.3). Over this area the potential drop over the PTCDI-$C_{13}H_{27}$/Alq_3 interface decreases with the distance from the insulator edge. Therefore, an OLED with field-effect electron transport has to be interpreted as the sum of sections of standard OLED devices, each with a different voltage over the PTCDI-$C_{13}H_{27}$/Alq_3 hetero-interface. This indeed leads to an overall weaker dependence of the light intensity on the pulse duration, as observed in Fig. 5.1.

5.2 Triplet Excitation Scavenging in Films of Conjugated Polymers

The accumulation of triplets is a problem inherent to organic materials and is a well-known issue also in classical liquid-state organic dye lasers, using highly fluorescent organic dye molecules dissolved in organic solvents. Triplet-state losses often limit the lasing performance of these dye lasers, especially at CW operation [296]. Consequently, organic dye lasers normally operate in pulsed mode (pulse duration of several nanoseconds) using a short-pulse flash-lamp for pumping since only in this way the accumulation of dye molecules in the triplet state can be overcome. To allow long-pulse operation, so-called triplet scavengers are usually employed [297]. This is described in Sect. 5.2.1. The use of triplet scavengers in solid-state organic thin films is investigated in Sect. 5.2.2.

5.2.1 Triplet Scavenging in Liquid-State Organic Dye Lasers

To circumvent triplet accumulation, introducing severe triplet-triplet absorption at the expected fluorescent lasing wavelength, the commonly accepted practice in liquid-state organic dye lasers is to use triplet scavengers [297]. It was demonstrated that the accumulation of dye molecules in the triplet state can be reduced by adding such molecules to the dye solution [297–299]. In this way, quenching of lasing emission by triplets might almost be eliminated. To reduce triplet accumulation, the triplet scavenger molecule should meet some important requirements:

- It should have the ability to accept a triplet excitation from the dye molecules, implying that its triplet level should be low enough.
- At the same time its singlet level should be high enough to prevent quenching of singlet excitations of the dye molecules. This implies thus that the S_1-T_1 splitting for the triplet scavenger should be extraordinary large.

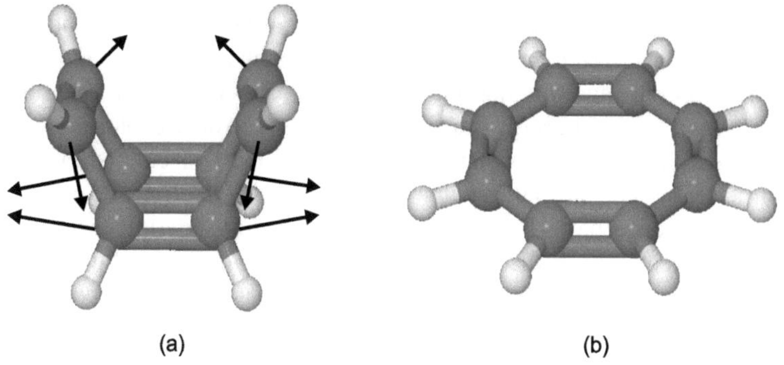

Fig. 5.5 (**a**) Illustration of the ground-state conformation of COT. The direction of the concerted C=C stretching and C–C torsional vibrations that planarize the molecule during triplet energy transfer are indicated by arrows. (**b**) The flat octagonal geometry of the COT triplet state

- It should possess a reasonably short intrinsic triplet lifetime to deplete quickly the triplet population or/and have intrinsic triplet-triplet absorption shifted far from the region of lasing of the dye molecules.
- It should not enhance intersystem crossing of the dye molecules to prevent converting singlets into triplet excitations. This criterion imposes a certain limitation for employing compounds containing heavy atoms (like metal-organic complexes) as triplet scavengers [300].

Taking into account all these requirements, it is not surprising that there are just a very limited number of efficient triplet scavengers available today.

The cyclic non-aromatic polyene, 1,3,5,7-cyclooctatetraene (COT), is the most popular and efficient triplet scavenger in liquid-state organic dye lasers due to its unique combination of properties, namely the ability to quench host triplets with an energy as small as 0.8 eV [301] and its very short triplet lifetime ($\sim$100 μs) [302]. The use of COT as an efficient triplet scavenger for several laser dye solutions is well documented [297–299], particularly, it was used commercially in Rhodamine-6G dye lasers to reduce excited triplet absorption. In addition, the mechanism for the energy-transfer process has been investigated in detail [301–303]. Unlike molecular oxygen, which is known as a notorious triplet quencher due to its triplet ground state, COT can quench triplets without increasing the intersystem crossing rate and without oxidation of the dye molecules. COT is a flexible molecule in the ground state and belongs to the "non-classical" triplet acceptors, which exhibit anomalous nonvertical (non-adiabatic) triplet energy transfer. The ground-state conformation of COT is illustrated in Fig. 5.5(a). During triplet energy transfer concerted C=C stretching and C–C torsional vibrations deform the molecule, favoring a more planar geometry as illustrated in Fig. 5.5(b) [301, 302].

Generally, an electronic transition is termed "vertical" with respect to the equilibrium geometry, conveying the idea that the electron is excited to the upper state before the nuclei have had the time to re-equilibrate. This is also known as an "adiabatic transition" [21]. It comes from the Franck-Condon principle (Sect. 1.1.3) according to which electronic transitions occur vertically since there is no change of the internuclear distance during the fast electron transition [20]. In other words, the nuclei are considered to be static with respect to the changes and movements of the electrons and the nuclear coordinates can be separated from the electronic part. It also implies that triplet energy transfer from a donor to an acceptor molecule normally is vertical, which means that the geometry of the molecule is not changing. This is true for organic molecules with rigid framework, which can be considered as "classical" triplet acceptors. For classical triplet acceptors the triplet energy transfer process depends mainly on the triplet energy of the molecule. This means that it is thermally activated in the case when the energy of the accepting molecule is higher than that of donor molecule (i.e. negative energy balance, $\Delta E_T < 0$) and that the activation energy is just the triplet energy difference [303].

"Nonvertical" transitions, on the other hand, imply a strong deformation of the acceptor molecule (also called "non-adiabatic transition") [304]. Non-vertical triplet energy transfer implies that the acceptor geometry changes concurrently with its excitation and therefore, it can only occur for some acceptor compounds, such as COT, cycloheptatriene [305], cis-stilbene, etc., which manifest significant flexibility. These molecules are also known as "non-classical" acceptors. In contrast to classical acceptors mentioned above, for nonvertical non-classical acceptors triplet energy transfer also depends on the molecule itself (its flexibility and geometrical distortion) and not only on its triplet energy [303]. Experimentally, nonvertical triplet acceptors exhibit a considerably faster triplet energy transfer rate in the regime of negative energy balance versus the vertical triplet energy (endothermic situation, $\Delta E_T < 0$), compared to the one predicted for a thermally activated Arrhenius process [304]. This is a consequence of the fact that nonvertical triplet energy transfer is accompanied by a geometrical distortion of the molecule, which results in a considerable lowering of the triplet state due to molecular relaxation. This relaxed triplet state of the nonvertical acceptor has considerably lower energy than that of the non-relaxed one. Consequently, nonvertical triplet acceptors can capture relatively low lying triplet excitations of donor molecules. The nonvertical acceptor COT, for example, can quench donor triplets with an energy $\geq$0.8 eV [301]. In addition, due to the relatively small energy distance between the relaxed triplet state and the ground state of such compounds, their triplet state possess a short lifetime as the probability of non-radiative energy dissipation increases exponentially with lowering the above energy difference [305].

5.2.2 *Triplet Scavenging in Solid-State Organic Thin Films*

Nonvertical triplet scavengers, such as COT, were so far only used in dye solutions and it is not obvious whether they can work in solid-state matrices as well. It is known that quenching of host triplets by COT requires a planarization of the COT molecule during triplet energy transfer [303, 305], otherwise the transition would be endothermic. In solution there are no steric restrictions for molecular deformation, however, it is not clear if such a structural reorganization could be hindered in a solid environment.

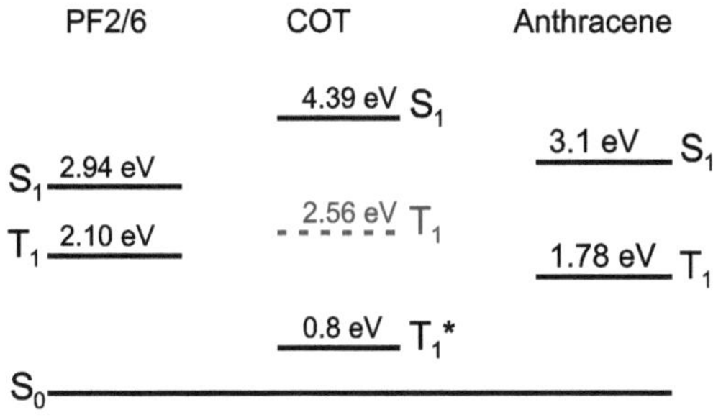

Fig. 5.6 Energy level scheme for PF2/6, COT and anthracene. The dashed line indicates the position of the T_1 triplet state of the COT molecule. This triplet state, however, is lowered during energy transfer due to molecular relaxation. The position of the relaxed triplet state of the COT molecule is indicated by T_1^*. (© Wiley-VCH Verlag GmbH & Co. KGaA. Reproduced with permission)

Besides, the energy transfer mechanism in liquid solutions is different from that in a solid-state matrix where it is dominated by a short-range Dexter mechanism (Sect. 1.3.3). When molecules are separated in liquid diluted solution, they have to approach each other to sufficiently small distance via diffusion of acceptor and donor molecules in the liquid in order to form an "encounter complex" where the triplet energy transfer process can happen. When the balance between the energies of the donor and acceptor triplet states (ΔE_T) is positive, triplet energy transfer in liquid solutions is controlled by diffusion of molecules in the liquid medium. In the case of negative energy balance ($\Delta E_T < 0$), the triplet energy transfer rate is additionally affected by a thermally activated Arrhenius process where ΔE_T is the activation energy. On the other hand, in solid-state organic molecules cannot easily diffuse and therefore the above mentioned molecular diffusion mechanism is not operative. Instead, triplet excitations can move from molecule to molecule via Dexter energy transfer owing to sufficiently small intermolecular separation (typically less than 1 nm). In other words, triplet energy transfer in solid matrices is controlled by diffusion of triplet excitons rather than by diffusion of molecules.

The goal of the present study is to elucidate whether "non-classical" triplet scavengers, such as COT, used before in liquid-phase organic dye lasers can also be used to scavenge host triplet excitations in solid-state highly fluorescent organic semiconductors. In our study we investigate the use of COT in polymer films of the polyfluorene derivative PF2/6. Polyfluorene is a prospective material for the fabrication of thin-film organic lasers [99, 141], where accumulation of triplets under electrical pumping is a very important issue. We also compare these results with those obtained by using a "classical" triplet acceptor with rigid molecular framework (anthracene), which also possesses suitable energetic levels to scavenge polyfluorene host triplets. The structural formulas of these compounds can be found in Chap. 2 (Sect. 2.1.1); the position of their S_1 and T_1 energy level are displayed in Fig. 5.6.

It should be mentioned that conventional vertical triplet energy transfer has been well documented for solid films of conjugated semiconducting polymers such as PPV derivatives [306], polyfluorenenes and MeLPPP [307], however, to the best of our knowledge no nonvertical triplet energy transfer has been reported for them.

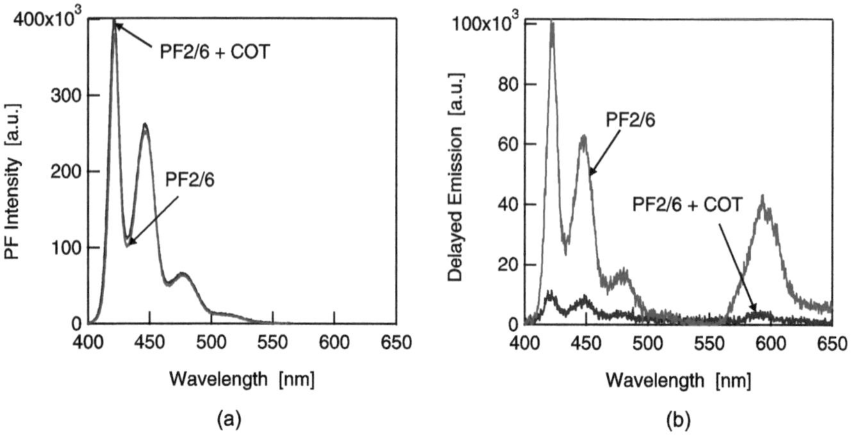

Fig. 5.7 (**a**) Prompt fluorescence intensity of a neat PF2/6 film and PF2/6 doped with 10 wt% COT registered at 77 K, and (**b**) Delayed emission spectra of the same films detected at 77 K at a time delay of 100 μs between the exciting laser pulse and the detection window. (© Wiley-VCH Verlag GmbH & Co. KGaA. Reproduced with permission)

Triplet Scavenging by the Non-classical Triplet Acceptor COT

Polymer films doped with triplet scavenger molecules were prepared by dissolving COT and PF2/6 in a ratio 1/10 in THF[1], and then drop-casting the resulting solutions on a precleaned quartz substrate. To delay evaporation of the COT, which is liquid at room temperature, the films were dried in a saturated COT environment. In order to exclude the presence of molecular oxygen in the fabricated organic films, solutions were prepared in a N_2-filled glovebox using de-aerated solvents.

Figure 5.7(a) shows the spectra of the prompt fluorescence (PF) from a neat PF2/6 film and from a PF2/6 film doped with 10 wt% COT registered at 77 K during the laser pulse excitation. These spectra are coincident with the steady-state PL spectra of these films, implying that the fluorescence in the range of 420–550 nm, caused by the radiative decay of singlet excitons, is dominant in this material. This is in line with earlier observations [308]. Note that virtually the same fluorescence intensity was observed in neat and doped PF2/6 films suggesting that the COT dopant does not quench singlet excitons in these films.

The delayed emission spectra of the same neat PF2/6 film and PF2/6 film doped with COT detected by introducing a time delay ($t_{del} = 100$ μs) between the exciting laser pulse and the detection window of the registration system is depicted in Fig. 5.7(b). As expected, delayed emission of the PF2/6 film consists of two different components with different spectral position. The short wavelength component in the blue spectral region, which is virtually identical to the prompt fluorescence spectrum, is due to delayed fluorescence (DF) from the S_1 state of PF2/6, while the second longer-wavelengths spectral component at ~590 nm can only be seen

[1] Tetrahydrofuran.

in the delayed emission and has been assigned to intrinsic phosphorescence (Ph) from the T_1 state [308]. It should be mentioned that, in contrast to very thin PF2/6 films (<100 nm) (not shown here), thicker drop-casted films normally do not exhibit vibronic fine structure of Ph spectra as could be seen from Fig. 5.7(b). This effect was explained before [308] in terms of somewhat different morphology of films of different thicknesses and agrees completely with the spectra reported earlier [308, 309]. Remarkable is that the intensity of the delayed emission from the neat and doped PF2/6 film was found to be drastically different, which is opposite to their prompt fluorescence (Fig. 5.7(a)). As can be seen on Fig. 5.7(b), both the DF and Ph are largely reduced in films doped with COT. This implies a strong quenching of the delayed emission in PF2/6 films by COT molecules, while at the same time the prompt fluorescence is not affected. This can be attributed to the extremely large singlet-triplet (S_1-T_1) splitting of the COT molecules (Fig. 5.6). The observed quenching effect is principally similar for different delay times.

A remarkable result of the present study is that it demonstrates that the non-classical triplet acceptor COT can be used as an efficient triplet scavenger also in a solid polymer film. Since during triplet energy transfer the COT molecule should be planarized [301–303, 305]—a process that can easily occur in liquid phase—the presented results suggest that such a structural reorganization seemingly is not hindered in a solid environment even at low temperature. In addition, the presented results prove that triplets' diffusion migration in the host PF2/6 polymer is sufficiently efficient even at low temperature. Therefore, triplets can be quenched similarly as in liquid solution, where triplet transfer occurs by dye-molecule colliding with the triplet scavenger.

Another important implication of this study is that it suggests that non-classical nonvertical triplet acceptors could be considered as a promising class of efficient triplet scavengers for solid-state electrically-pumped organic light-emitting devices due to (i) an exceptionally large energy splitting between the lowest excited singlet state (4.39 eV for COT) [302] and the relaxed triplet state, which is inherent for these compounds, and (ii) a reasonably short triplet lifetime (~100 μs for COT) that enables fast deactivation of triplet excitations. In this way, the accumulation of triplets, introducing triplet-state losses such as singlet-triplet annihilation and triplet-triplet absorption, might be considerably reduced.

Triplet Scavenging by the Classical Triplet Acceptor Anthracene

To compare the observed host triplet quenching effect in PF2/6 due to "non-classical" COT dopants with the effect of the presence of "classical" triplet scavengers, thin PF2/6 films doped with anthracene were prepared. The conventional aromatic molecule anthracene (energy levels shown in Fig. 5.6) has a very low triplet level (1.78 eV) [9] and an exceptionally large S_1-T_1 energy level splitting (1.32 eV) compared to other organic conjugated small molecules and was therefore chosen as triplet acceptor. Anthracene additives in PF2/6 films present thus an example of a classical triplet scavenger for which a vertical (adiabatic) triplet energy transfer

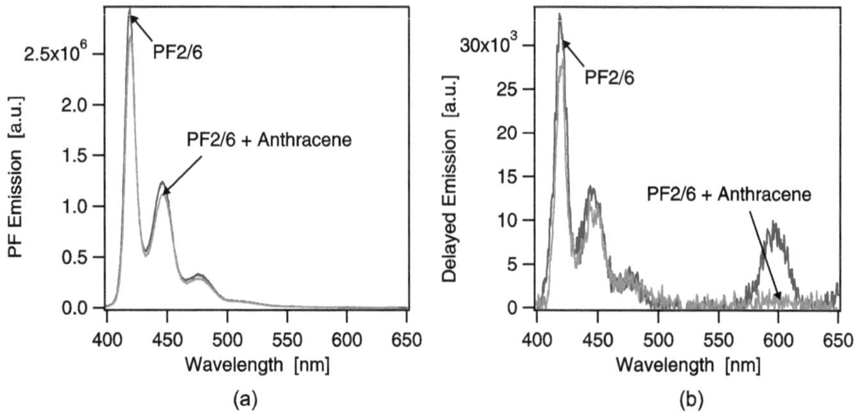

Fig. 5.8 (**a**) Prompt fluorescence, and (**b**) delayed emission spectra of a neat PF2/6 film and a PF2/6 film doped with 20 wt% anthracene. The spectra were measured at 77 K at a time delay of 100 μs. (© Wiley-VCH Verlag GmbH & Co. KGaA. Reproduced with permission)

process is operative. Anthracene doped PF2/6 films were prepared by spin-coating a 20 wt% solution from toluene at 1000 rpm. Also here, solutions were prepared in a N_2-filled glovebox using de-aerated solvents in order to exclude the presence of molecular oxygen in the organic films. A comparison of the prompt fluorescence of the neat PF2/6 and PF2/6 film doped with anthracene at 77 K is presented in Fig. 5.8(a). Delayed emission spectra of these films ($t_{del} = 100$ μs) are shown in Fig. 5.8(b).

As can be seen from Fig. 5.8(b), anthracene dopants in PF2/6 films do quench host phosphorescence due to exothermic triplet energy transfer from the triplet state of the polymer ($E_T^{PF2/6} = 2.10$ eV) [308] to the one of anthracene ($E_T^{anthracene} =$ 1.78 eV) [9] and this quenching effect is not very different from the one observed with the non-classical COT triplet acceptor. Since the singlet state of anthracene (3.1 eV) [9] is higher than the one of PF2/6, it is natural that the prompt fluorescence intensity of the polymer is not quenched by the presence of anthracene additives (Fig. 5.8(a)).

Comparison of the delayed emission spectra of PF2/6 doped with COT and anthracene additives shows, however, that these triplet scavengers have very different effect on the delayed fluorescence (DF) of the PF2/6 polymer. As can be seen, while COT dopants strongly quench the DF of the PF2/6 (Fig. 5.7(b)), anthracene doping has almost no effect on the host DF (Fig. 5.8(b)).

To rationalize these observations one should consider the origin of the DF in PF2/6 polymer films. Previous time-resolved PL studies [308] reveal that the DF emission in PF2/6 polymer films is dominated by triplet-triplet annihilation. Triplet-triplet annihilation arises due to a bimolecular reaction between two triplet excitations ($T_1 + T_1 \rightarrow S^* + S_0$) and therefore, the DF intensity should be a quadratic of the triplet concentration. Our measurements support such a notion as indeed the DF intensity in the PF2/6 film was found to increase quadratically with the excitation intensity within the considered laser intensity range. This is illustrated in Fig. 5.9(a),

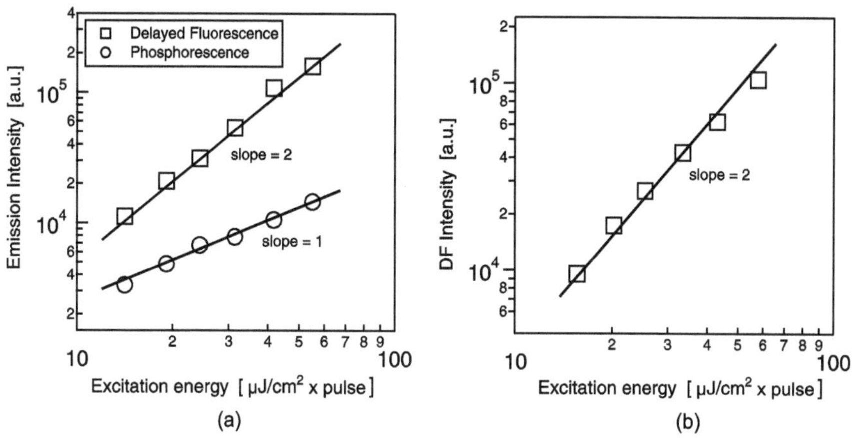

Fig. 5.9 Intensity dependence upon laser excitation intensity of (**a**) the DF and Ph emission in neat PF2/6 films, and (**b**) the DF emission in PF2/6 films doped with anthracene. (© Wiley-VCH Verlag GmbH & Co. KGaA. Reproduced with permission)

which shows the intensity dependence of the DF and Ph emissions in neat PF2/6 films upon the laser excitation intensity. To have the maximum signal possible, the delayed emission spectra were recorded with a time delay of 75 ns and a gate width of 10 ms, i.e. in integration mode. The emission intensities plotted in Fig. 5.9(a) are the peak emission intensities (0-0 peak for the DF). The quadratical increase of the DF intensity upon laser excitation clearly points to a bimolecular origin of the DF emission (DF $\sim I_{exc}^2$). At the same time the phosphorescence intensity of the polymer increases linearly with the pumping intensity (Ph $\sim I_{exc}$), implying that the DF is quadratically proportional to the triplet concentration in the film.

Reduction of the DF intensity in PF2/6 films doped with COT can be readily explained by the fact that the COT molecules capture host triplets and deactivate them within the relatively short lifetime of the COT triplet state ($\sim$100 μs), thus preventing their possible annihilation. Anthracene additives in PF2/6 films also capture host triplets efficiently as evidenced by a strong reduction of the host phosphorescence (Fig. 5.8(b)); however, due to the very large radiative lifetime of the anthracene triplet state ($>$10 ms) [9] triplets are not deactivated quickly, but rather accumulated on the anthracene molecules. The results above clearly indicate that in order to control the triplet concentration efficiently, a suitable triplet scavenger should not only possess a sufficiently large S_1-T_1 energy level splitting, but should also have a reasonably short triplet lifetime, making a fast deactivation of triplet excitations possible. Anthracene additives are therefore not useful to reduce the triplet concentration. They may scavenge host triplet excitations, but due to their long triplet lifetime an accumulation instead of a reduction of the triplet population will occur.

Furthermore, the DF emission in PF2/6 films doped with anthracene was found to originate also from a bimolecular process in the host-guest system since the DF intensity increases quadratically with the excitation intensity as can be seen from Fig. 5.9(b). Since the host triplets are eventually transferred to the anthracene guest, as proven by strong quenching of the intrinsic PF2/6 phosphorescence (Fig. 5.8(b)), T_{host}-T_{host} annihilation cannot be responsible for the observed DF. On the other hand, T_{guest}-T_{guest} annihilation is also unlikely to be a dominant mechanism for the DF owing to the dopant dilution in the polymer host that precludes short-range Dexter-type triplet migration via the guest manifold. We therefore assume that the observed bimolecular annihilation proceeds via a T_{guest}-T_{host} annihilation reaction, i.e., when a mobile host triplet encounters a long-lived localized guest triplet excitation on an anthracene molecule, although clarifying this issue requires further investigations.

It has to be noticed that a similar mechanism was suggested before to explain DF in some conjugated polymers doped with metal-organic complexes [310, 311]. During such a T_{guest}-T_{host} annihilation process the excited singlet state of either the host or the guest molecule can be populated with equal probability, but since the host singlet state is lower than the singlet state of the anthracene dopant, only DF fluorescence of the PF2/6 polymer is observed.

5.3 Summary and Conclusions

In this chapter, two different approaches to reduce the triplet concentration in organic light-emitting devices were investigated. First, the pulsed excitation behavior of OLEDs with field-effect-assisted electron transport was discussed and their performance was compared to conventional OLEDs using the same organic layers. We found that not charge carrier transport, but injection of electrons from the electron-transporting layer into the active light-emitting layer determines the dependence of the light intensity on the applied pulse width. Since this interface is present in conventional OLEDs as well as in OLEDs with field-effect electron transport, similar pulse-width dependence was observed for both devices. Furthermore, we proved that, depending on the applied voltage and for this particular material combination, pulses down to 1 μs could be applied without affecting the light intensity. This fact is important for the control of the triplet concentration. 1 μs is much shorter than the typical triplet lifetime of fluorescent materials, indicating that pulsed excitation could be applied to reduce triplet accumulation in our device.

An alternative way to reduce the triplet concentration is the incorporation of a triplet scavenger, as discussed in the second part of this chapter. Phosphorescence and delayed fluorescence of polyfluorene polymer films doped with cyclooctatetraene (COT) and anthracene were investigated by time-resolved photoluminescence measurements. Using the "non classical" triplet scavenger COT, employed before in liquid-phase organic dye lasers, anomalous nonvertical triplet energy transfer in solid conjugated polymer films was demonstrated for the first time. This triplet energy transfer behaves similarly to the one of vertical triplet acceptors like anthracene. Thanks to their large singlet-triplet splitting, both dopant molecules were found to quench efficiently host phosphorescence without affecting the host fluorescence. This S_1-T_1 energy level splitting is exceptionally large in COT due to its low-lying relaxed triplet state capable of accepting host triplet excitations. In addition, contrary to anthracene, the triplet lifetime of COT molecules is reasonably short, making a fast deactivation of triplet excitations possible. This suggests that nonvertical triplet scavengers, which possess a large S_1-T_1 energy level splitting and an inherently short lifetime of their relaxed triplet state, could be promising candidates for quenching host triplet excitations in electrically pumped organic lasers and other high-brightness organic light sources.

Chapter 6
Triplet-Emitter Doped Organic Materials

OLEDs have been extensively investigated for full-color, plat-panel display applications and illumination light sources because of their merits of high luminance, low driving voltage, and variety of emission colors [29]. The internal quantum efficiency of electro-fluorescence OLEDs is limited to around 25% according to simple spin statistics [36]. In order to overcome this efficiency limit of fluorescent OLEDs, phosphorescent emitters doped into host material have been used [312]. Phosphorescent emitters are low-molecular weight phosphorescent dye molecules incorporating a heavy metal atom with strong spin-orbit coupling that mixes singlet and triplet states. The radiative decay of triplet states becomes allowed and the efficiency of intersystem crossing is enhanced. As a result, the lowest triplet state is efficiently populated and produces light emission with a large quantum yield. Phosphorescent emitters in both small organic molecule hosts and polymer hosts allow for harvesting singlet and triplet excitons and, therefore, the internal quantum efficiency of OLEDs can be largely enhanced, approaching 100% [34].

In this chapter, triplet-emitter doped organic materials are studied. Section 6.1 focuses on the triplet dynamics and charge carrier trapping in triplet-emitter doped conjugated polymers. Phosphorescent properties and charge carrier trapping are studied by time-resolved photoluminescence and thermally-stimulated luminescence techniques. In Sect. 6.2, the use of triplet-emitters as lasing element is investigated. The results of Sect. 6.1 have been published in [311].

6.1 Triplet Dynamics and Charge Carrier Trapping in Triplet-Emitter Doped Conjugated Polymers

For the preparation of electro-phosphorescent polymer light-emitting diodes, the commonly used concept is to blend a low-molecular weight phosphorescent emitter into a proper polymer matrix. Transfer of excited state energy plays an important

S. Schols, *Device Architecture and Materials for Organic Light-Emitting Devices*,
DOI 10.1007/978-94-007-1608-7_6, © Springer Science+Business Media B.V. 2011

role in the operation of these devices. Singlet and triplet excitons can first be generated by electron-hole recombination in the polymer host and then transfer their energy to the dopant. Alternatively, the triplet state of the phosphorescent emitter may be formed by charge transfer from the host followed by charge recombination at the guest molecule. In both cases, the singlet and triplet energy levels of the host have to be well above the triplet state of the guest emitter to be efficient and prevent back transfer. The actual transfer of excitation to the guest molecule can occur via different mechanisms: Förster transfer of singlet excitons generated on the polymer matrix to the guest, Dexter transfer of both singlet and triplet excitons generated on the host to the dopant, as well as the direct generation of singlet and triplet excitons on the guest [313, 314].

Comprehensive studies have been performed to investigate triplet dynamics in MeLPPP doped by PtOEP[1] [307, 310, 315]. Since both the S_1 and T_1 levels of the dopant are below those of the matrix [316–318], both Förster-type host-guest singlet-singlet and Dexter-type triplet-triplet energy transfer can be envisaged. A specific feature of devices based on phosphorescence is that long phosphorescent lifetimes lead to saturation of emissive sites, and annihilation of triplet excitations may cause significant efficiency losses. Triplet-triplet annihilation is believed to be one of the major quenching mechanisms in phosphorescent OLEDs [266]. As a result, the efficiency of these OLEDs tends to decrease at high current densities [319]. The longer lifetime of triplet excitons also reduces one of the advantages of dye doping of organic light-emitting devices, namely the ability to eliminate concentration quenching. One way to solve these problems is to use dopants with shorter triplet exciton lifetime. Therefore Ir-complexes are more attractive candidates than platinum porphyrins owing to about one order of magnitude shorter lifetimes, which typically range from 1 to 14 μs [260].

In this section, we explore the phosphorescent properties of the red light-emitting metal-organic complex $Btp_2Ir(acac)$ doped into the conjugated polymer CNPPP. Similar to the above mentioned MeLPPP:PtOEP host-guest system, both the S_1 and T_1 levels of the $Btp_2Ir(acac)$ dopant are below those of the CNPPP polymer matrix; however, the phosphorescent lifetime of $Btp_2Ir(acac)$ (~7 μs [320]) is tenfold shorter than that of PtOEP (~80 μs [321]). It will be demonstrated that this results in different mechanisms dominating the phosphorescence properties. In addition, direct evidence that the phosphorescent metal-organic complexes do act as charge traps is presented and these traps are characterized by means of charge trapping spectroscopy. Polymer films doped with triplet-emitter molecules were prepared by dissolving the appropriate ratios of $Btp_2Ir(acac)$ and CNPPP in toluene, and then spin-coating the resulting solutions on a precleaned quartz substrate. The structural formulas of both compounds and the specific processing conditions used to prepare the organic thin films can be found in Chap. 2 (Sect. 2.1.1).

[1] Platinum(II) octaethyl porphyrin.

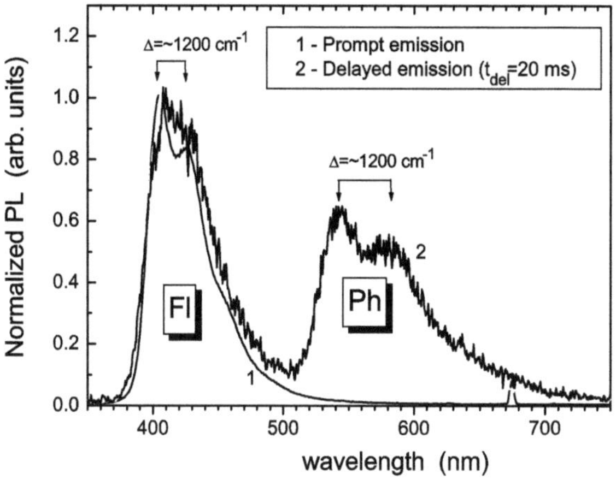

Fig. 6.1 Normalized spectra of prompt fluorescence (curve 1) and delayed emission (curve 2) from a pristine CNPPP film at 80 K. Delayed emission was measured with 20 ms delay time after the laser pulse and with a gate width of 6 ms. (© Elsevier)

6.1.1 Spectroscopic Characterization

Pristine CNPPP Films

Figure 6.1 (curve 1) shows the prompt fluorescence spectrum of a pristine CNPPP film registered during the laser pulse at 80 K. The spectrum consists of a main $S_1 \rightarrow S_0$ band at 405 nm (3.08 eV) followed by a vibronic feature displaced by about 1200 cm^{-1}. This spectrum coincides with steady-state PL spectra, implying that the fluorescence in the range of 400–500 nm, caused by the radiative decay of singlet excitons, is dominant in this material.

The delayed emission spectrum of the pristine CNPPP film, detected by introducing a time delay ($t_{del} = 20$ ms) between the exciting laser pulse and the detection window of the registration system, is depicted by curve 2 in Fig. 6.1. In contrast to prompt fluorescence, the delayed emission from CNPPP exhibits two different components with different decay times and different spectral positions. The short wavelength component of the delayed emission is virtually identical to the prompt fluorescence spectrum. Consequently, it can be assigned to delayed fluorescence (DF) from the S_1 state of CNPPP. The second, slower decaying set of bands is shifted by 0.78 eV with respect to the first peak of the DF, but has the same vibronic splitting of about 1200 cm^{-1} (Fig. 6.1, curve 2) as the fluorescence spectrum. It should be noted that this second longer-wavelengths spectral component can only be seen in the delayed emission and shows all generic signatures of phosphorescence, principally similar to that previously found in other conjugated polymers as MeLPPP [322, 323], LPPPC[2] [324] and polyfluorenes [308]. Therefore, this spectral part in the range of 520–700 nm (Fig. 6.1, curve 2) is assigned to intrinsic phosphorescence (Ph) from the T_1 state at 542 nm (2.3 eV). Pristine CNPPP thus features a singlet-triplet (S_1-T_1) energy level splitting of about $\Delta E_{ST} = 0.78$ eV, which is a typical value for organic conjugated polymers [306].

[2]Ladder-type poly(para-phenylene carbazole).

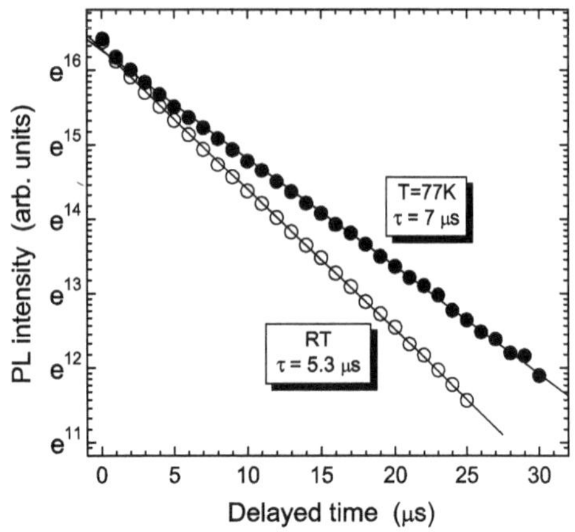

Fig. 6.2 Time decay kinetics of Ph emission measured in a PS:(1 wt%)$Btp_2Ir(acac)$ film at 77 K and room temperature. The gate width was 0.1 μs. (© Elsevier)

$Btp_2Ir(acac)$ Doped into a PS Matrix

In order to get an insight into the intrinsic Ph properties of $Btp_2Ir(acac)$, the CW- and time-resolved PL of this triplet-emitter doped into a polystyrene (PS) film has been studied. Because PS did not show any light emission under the excitations used in this study, it can be considered as an optically inert polymer host allowing matrix isolation of $Btp_2Ir(acac)$ molecules. Therefore, Ph properties of PS:$Btp_2Ir(acac)$ films are virtually the same as those measured in frozen diluted solution. The Ph decay kinetics of a $Btp_2Ir(acac)$ (1 wt%) doped PS film recorded at 77 K and room temperature are depicted in Fig. 6.2. The Ph decay curves demonstrate a clear mono-exponential kinetics, indicating that the triplet excitations of $Btp_2Ir(acac)$ are efficiently confined by the PS host at such dopant concentration and that triplet quenching by impurities or via T-T-annihilation mechanism is negligible. The intrinsic lifetime of Ph emission decreases from 7 μs at 77 K to about 5 μs when the temperature increases to room temperature. This correlates with the observed gradual decrease of the Ph intensity of $Btp_2Ir(acac)$ with elevating temperature (Fig. 6.3) and suggests that temperature-dependent non-radiative processes are active in this metal-organic complex [320].

Figure 6.4 presents the low-temperature Ph decay kinetics of $Btp_2Ir(acac)$ doped PS films at small and relatively high dopant concentration (1 and 10 wt%). As can be seen, the apparent Ph lifetime is notably smaller ($\tau = 5.3$ μs) at 10 wt% $Btp_2Ir(acac)$ concentration compared to a concentration of 1 wt% ($\tau = 7$ μs). This quenching is definitely related to the dopant concentration (concentration quenching) and not to T-T annihilation as verified by the linear intensity dependence of Ph in PS:(10 wt%)$Btp_2Ir(acac)$ upon excitation intensity (inset of Fig. 6.4).

Figure 6.5 shows the profile of the first Ph band ($T_1 \rightarrow S_0$) from $Btp_2Ir(acac)$ in a frozen diluted toluene solution ($c = 10^{-3}$ wt%) and from $Btp_2Ir(acac)$ doped in a PS film at $c = 10$ wt% measured at different delay time (Fig. 6.5(a) and (b), respectively). The Ph peak position of $Btp_2Ir(acac)$ in frozen solution showed no peak shift (Fig. 6.5(a)), whereas in heavily doped PS films, a gradual red-shift of the peak with increasing delay time (Fig. 6.5(b)) was observed. The same red-shift was

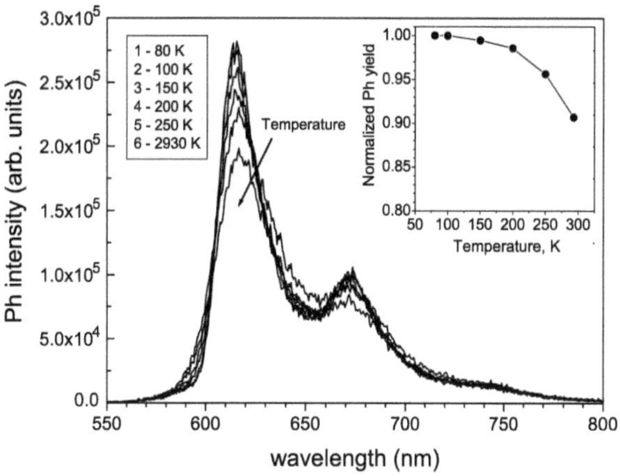

Fig. 6.3 Ph intensity of a 1 wt% $Btp_2Ir(acac)$ doped PS film measured at different temperatures. At elevated temperatures a decrease of the Ph yield is observed. (© Elsevier)

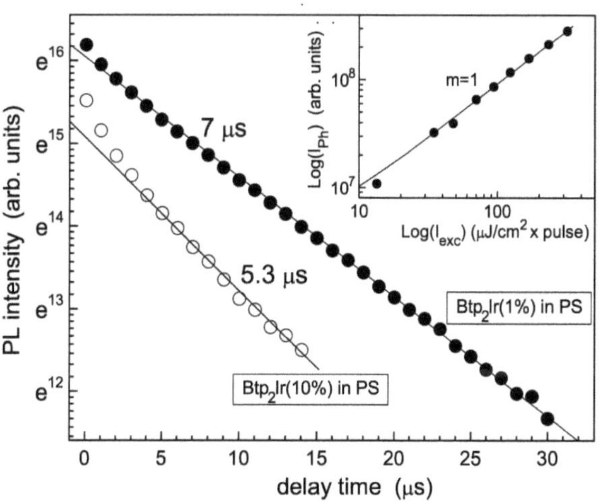

Fig. 6.4 Time decay kinetics of Ph emission measured in PS:$Btp_2Ir(acac)$ films at 1 and 10 wt% concentrations of $Btp_2Ir(acac)$. The kinetics were measured at 80 K with a gate width of 0.1 μs. *Inset*: dependence of the Ph intensity in a PS:(10 wt%) $Btp_2Ir(acac)$ film upon the laser excitation intensity. (© Elsevier)

observed for the vibronic replica peaks (not shown here). This shift is a direct observation of the so-called spectral diffusion effect of the Ph spectra of $Btp_2Ir(acac)$ and is the result of triplet migration from one excited state to another during the triplet lifetime. Since triplet diffusion towards excited states with a lower energy is favored (downward hopping), a red-shift of the Ph peak is observed. This process is schematically illustrated in Fig. 6.6.

The observation of spectral diffusion proves that triplets can migrate through the $Btp_2Ir(acac)$ molecules manifold at a doping level of 10 wt%. It actually implies that triplets are no longer confined on isolated $Btp_2Ir(acac)$ molecules at such doping concentration, which is close to that often used for fabrication of high-efficiency phosphorescent OLEDs [321]. Consequently, at such a concentration, the triplets could be amenable to quenching. Indeed, minor traces of quenching impurities, such as triplet molecular oxygen or some intrinsic quenching defects in the PS polymer,

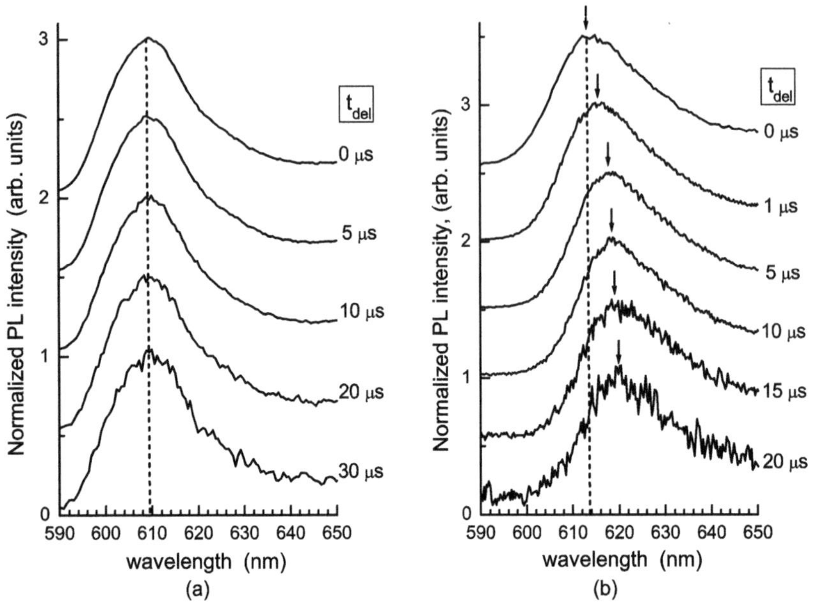

Fig. 6.5 (**a**) Time resolved first Ph peak from $Btp_2Ir(acac)$ in a frozen diluted toluene solution ($c = 10^{-3}$ wt%), and (**b**) Time resolved first Ph peak from $Btp_2Ir(acac)$ doped in a PS film at $c = 10$ wt%. All spectra were measured at 80 K and normalized to the peak intensity. (© Elsevier)

Fig. 6.6 Schematical illustration of the spectral diffusion effect. (© Elsevier)

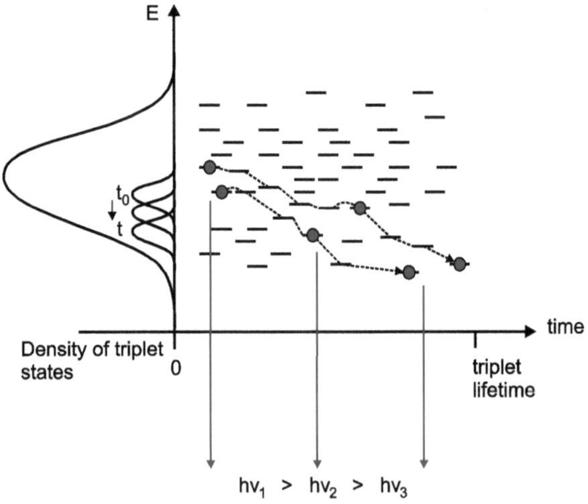

could become accessible to migrating triplets within the radiative lifetime of the triplet excitation [323].

In general, short-range Dexter energy transfer dominates migration of triplet excitations. The observation of triplet migration at a $Btp_2Ir(acac)$ concentration of 10 wt% is, however, in good agreement with the recently proposed mechanism of dipole-dipole long-range Förster energy transfer in phosphorescent Ir(III) complexes [325]. Kawamura *et al.* found out that the strong spin-orbital coupling in these compounds increases the oscillator strength of the triplet-singlet transition and leads to a sufficiently high dipole-dipole transfer rate to enable Förster energy transfer to be the dominant mechanism in these phosphorescent Ir-complexes [325]. An effective Förster radius of 1.5 nm was calculated for $Btp_2Ir(acac)$ [325]. The average distance between the $Btp_2Ir(acac)$ molecules in a PS host at a doping concentration of 10 wt% is $\rho = 2.2$ nm. This value was calculated using the so-called lattice gas model [326]:

$$\rho = \left(\frac{M}{N_A c \delta}\right)^{1/3}, \tag{6.1}$$

where M is the $Btp_2Ir(acac)$ molecular weight, N_A is the Avogadro's constant, c is the fractional $Btp_2Ir(acac)$ concentration and δ is the sample density. Taking into account that Eq. 6.1 gives the average distance between the center of the molecules and thus slightly overestimates the intermolecular distance, 2.2 nm, determined above, is not very different from the effective Förster radius obtained by Kawamura. Förster energy transfer might thus indeed explain the spectral diffusion effect observed in the present study at 10 wt% dopant concentration (Fig. 6.5(b)).

The results presented in Fig. 6.5 elucidate the observed apparent decrease of the exciton lifetime of $Btp_2Ir(acac)$ at increased dopant concentration in PS films (Fig. 6.4). If triplet excitons are no longer confined on individual $Btp_2Ir(acac)$ molecules at 10 wt% dopant concentration, as proven by the observed spectral diffusion effect, the probability of triplet quenching by accidental impurities, which is suppressed in a highly diluted system, increases with concentration, giving rise to the observed concentration quenching effect. The quenching impurity is most probably triplet molecular oxygen, which is a notoriously efficient triplet exciton quencher. Comparison of the Ph decay kinetics in oxygen-free encapsulated films and that of films exposed to the ambient atmosphere strongly supports such a notion. Figure 6.7 shows the low-temperature Ph decay kinetics of PS:$Btp_2Ir(acac)$ films at high (10 wt%) triplet-emitter concentration. As can be seen, oxygen-free (never exposed to air and encapsulated in the glove-box) films (Fig. 6.7, curve 1) show the same mono-exponential Ph decay with a lifetime of 7 μs as for diluted systems, suggesting negligible triplet quenching in them. On the other hand, PS:(10 wt%)$Btp_2Ir(acac)$ films exposed to air feature an initial fast non-exponential decay (curve 2), which is a clear indication of quenching by oxygen. In combination with the results presented in Fig. 6.5, this suggest straightforwardly that the concentration quenching in PS:$Btp_2Ir(acac)$ systems arises due to triplet exciton migration at increased guest molecule concentration and thus due to their ability to encounter non-radiative oxygen impurities.

It should be mentioned that a significant phosphorescence lifetime shortening due to oxygen was recently reported [307] for another host-guest system, namely MeLPPP doped with PtOEP (concentration from 10^{-3} to 5 wt%). The effect was explained by reversible formation of a weak complex between PtOEP and oxygen upon exposing the sample to air, which quenches the phosphorescence of PtOEP. It

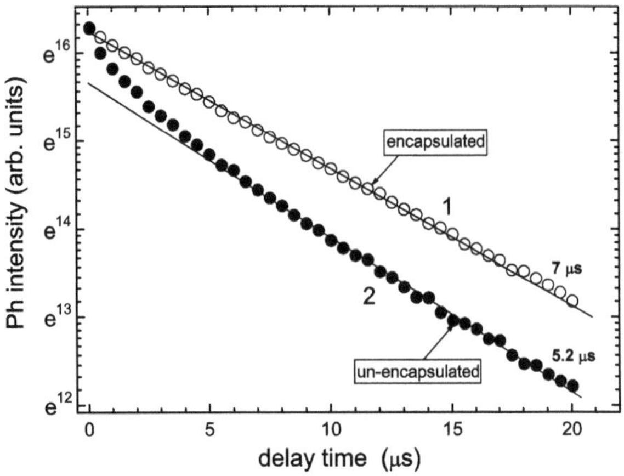

Fig. 6.7 Low-temperature Ph decay kinetics of a PS:(10 wt%)$Btp_2Ir(acac)$ film encapsulated between two quartz plates in the glovebox (curve 1) and for a film prepared from the same solution but exposed to the air before loading to the cryostat (curve 2). (© Elsevier)

was assumed that not mobile but localized guest triplets are involved in the quenching effect [307]. In our study, however, it is observed that the oxygen quenching effect on the phosphorescence of $Btp_2Ir(acac)$ depends strongly on the guest concentration (Fig. 6.4), as no notable lifetime shortening was found at 1 wt% $Btp_2Ir(acac)$ concentration either in a PS (Fig. 6.4) or CNPPP (see further down, Fig. 6.9(b)) host matrix, while at a larger (10 wt%) dopant concentration the quenching effect was obvious. The remarkable correlation of the quenching effect with the onset of the spectral diffusion of the phosphorescence of $Btp_2Ir(acac)$ dispersed in a polymer matrix proves that in the present experiment, using $Btp_2Ir(acac)$ dopants, mobile triplets are involved and that they are able to reach the oxygen quenching center at sufficiently large dopant concentration. This discrepancy might be rationalized by taking into account the considerably shorter intrinsic phosphorescence lifetime of $Btp_2Ir(acac)$ (7 μs vs. ~80 μs for PtOEP) and, consequently, the somewhat larger oscillator strength of the radiative transition in this organometallic complex as compared to PtOEP molecule.

Finally, it should be noted that the presented results suggest that the concentration quenching effect, at least at 10 wt% $Btp_2Ir(acac)$ concentration level, is definitely due to quenching by impurities like molecular oxygen and not due to the mechanism based on dampening of energy through multiple dipole-dipole energy exchange interaction between neighboring Ir-complex molecules as was suggested before [325]. Indeed, Fig. 6.5(b) proves that triplet energy migration does occur at this dopant concentration, but oxygen-free encapsulated PS:(10 wt%)$Btp_2Ir(acac)$ films do not show any shortening of the triplet lifetime (Fig. 6.7). An energy dampening mechanism [325, 327] may, however, still be operative at smaller average distances between $Btp_2Ir(acac)$ molecules (i.e. at larger dopant concentrations).

CNPPP Films Doped with $Btp_2Ir(acac)$

Figure 6.8(a) shows the normalized steady-state low-temperature PL spectra of CNPPP films doped with $Btp_2Ir(acac)$ at different doping concentrations. As can

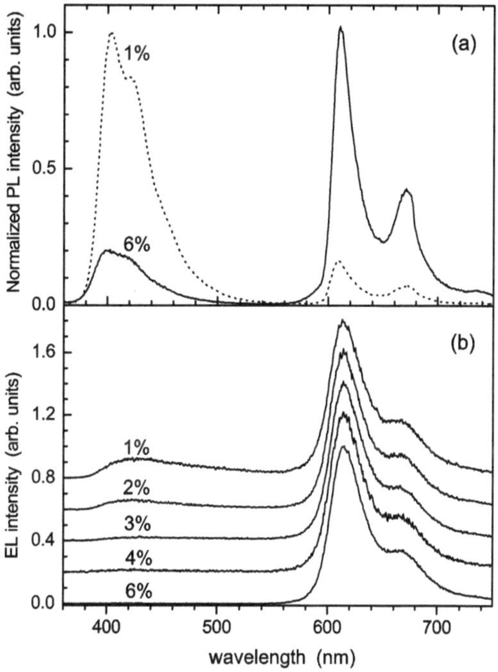

Fig. 6.8 (**a**) Steady-state PL spectra of CNPPP films doped with $Btp_2Ir(acac)$ at doping concentration of 1 and 6 wt% measured at 80 K, and (**b**) EL spectra of CNPPP:$Btp_2Ir(acac)$ films with doping concentration varying from 1 to 6%. (© Elsevier)

be seen, the PL spectra exhibit two components corresponding, respectively, to fluorescent emission from the CNPPP host in the range from 400 to 500 nm, and a set of red emission peaks of $Btp_2Ir(acac)$ phosphorescence. The phosphorescence (Ph) of $Btp_2Ir(acac)$ shows a well-resolved spectrum with the $T_1 \rightarrow S_0$ band at 610 nm (2.05 eV), followed by a vibronic fine structure. The mechanism of Ph in this compound has been assigned before to the transition from ligand-based triplet $^3(\pi - \pi^*)$ states [260, 321]. Time-resolved PL studies of pure CNPPP films (Fig. 6.1, curve 2) have revealed the intrinsic phosphorescence in this polymer and allow direct estimation of the triplet energy of 2.3 eV in this material. Since this CNPPP triplet level exceeds the guest triplet level at 2.05 eV, efficient confinement of $Btp_2Ir(acac)$ triplet excitations is possible in the CNPPP:$Btp_2Ir(acac)$ host-guest system. Clearly, the relative contribution from the fluorescence emission of the CNPPP host reduces significantly with increasing $Btp_2Ir(acac)$ concentration. This is expected due to Förster singlet energy transfer from the host matrix to the dopant, followed by intersystem crossing to the triplet state of the iridium complex [328]. However, even at dopant concentrations as high as 6 wt% the host fluorescence is not completely quenched (Fig. 6.8(a)). Note that light absorption at the excitation wavelength $\lambda_{exc} = 337$ nm in CNPPP:$Btp_2Ir(acac)$ thin films is dominated by CNPPP host absorption at all considered dopant concentrations. Therefore, excitation of $Btp_2Ir(acac)$ molecules could only be possible via energy transfer from the host matrix.

The electroluminescence (EL) spectra of $Btp_2Ir(acac)$ doped CNPPP films are shown in Fig. 6.8(b). These spectra were measured at different $Btp_2Ir(acac)$ doping concentration and by using an ITO/PEDOT/CNPPP:$Btp_2Ir(acac)$/LiF/Al device configuration. As can be seen, in contrast to the PL emission spectra (Fig. 6.8(a)), a complete quenching of the host EL emission occurs at notably lower $Btp_2Ir(acac)$ doping concentration (as low as 3 wt%). EL quenching commonly occurs at much lower triplet emitter concentration as compared to the PL emission and has been attributed by many authors to charge recombination taking place preferentially on triplet emitter molecules, tacitly assuming efficient charge-carrier trapping by the dopant species [328, 329].

The intensity dependence of prompt fluorescent and Ph emission of CNPPP films doped with 1 wt% $Btp_2Ir(acac)$ upon laser excitation intensity is presented in Fig. 6.9(a). The plotted emission intensities were obtained by integration over the wavelength. The CNPPP fluorescence intensity varies approximately linearly with the laser power (curve 1). The Ph of $Btp_2Ir(acac)$, on the other hand, behaves in a different way (Fig. 6.9(a), curve 2): at low excitation intensity the dependence of Ph is a linear function ($I_{ph} \sim I_{exc}^{m}$, where $m = 1$), whereas at power excitation densities higher than ~50 μJ/(cm^2 pulse) a crossover to a slow square-root increase ($m = 0.5$) occurs. Such intensity dependence of Ph at high excitation densities usually arises from a bimolecular quenching process which competes with the monomolecular decay and is often considered as a signature of triplet-triplet annihilation (T-T-annihilation) ($T_1 + T_1 \rightarrow S^* + S_0$) [308, 323], implying a decrease of the Ph yield at high pump fluence. The slow square-root increase with increasing excitation intensity cannot be related to the host singlet state since the fluorescence of this host-guest system does not show saturation behavior (Fig. 6.9(a), curve 1). In addition, it was found that the excitation intensity at which the above crossover is observed shifts to somewhat lower intensities with increasing temperature, as expected for a thermally activated diffusion migration of triplet excitations governing the T-T annihilation.

Figure 6.9(b) shows the decay kinetics of phosphorescent emission of $Btp_2Ir(acac)$ in the same CNPPP:(1 wt%)$Btp_2Ir(acac)$ film as in Fig. 6.9(a) measured at low- and high- laser excitation intensity at 80 K (curve "a" and "b", respectively). The excitation intensities used are indicated by arrows in Fig. 6.9(a) and these laser powers correspond to the regime of either a linear- or a square-root-Ph increase with increasing pump intensity. At low excitation intensity the Ph of $Btp_2Ir(acac)$ doped in a CNPPP matrix decays mono-exponentially over the studied time range, featuring a lifetime of about $\tau = 7$ μs (Fig. 6.9(b), curve "a"). This observation, which is virtually the same as has been observed in an optically inert PS host (Fig. 6.9(b), open symbols) at 1 wt% dopant concentration, indicates that the CNPPP host indeed can efficiently confine triplet excitations of $Btp_2Ir(acac)$ molecules. At high excitation intensity (Fig. 6.9(b), curve "b"), on the other hand, the decay kinetics deviate from a single exponent; a fast initial Ph decay during several microseconds after the laser pulse is followed by a mono-exponential decay with the same lifetime of 7 μs at larger delay times. This behavior proves that efficient bimolecular T-T annihilation occurs during the first 10 μs after the laser pulse when the concentration of triplets in the material is still large enough. T-T annihilation leads to the depletion of the reservoir of triplet excitations and therefore the de-

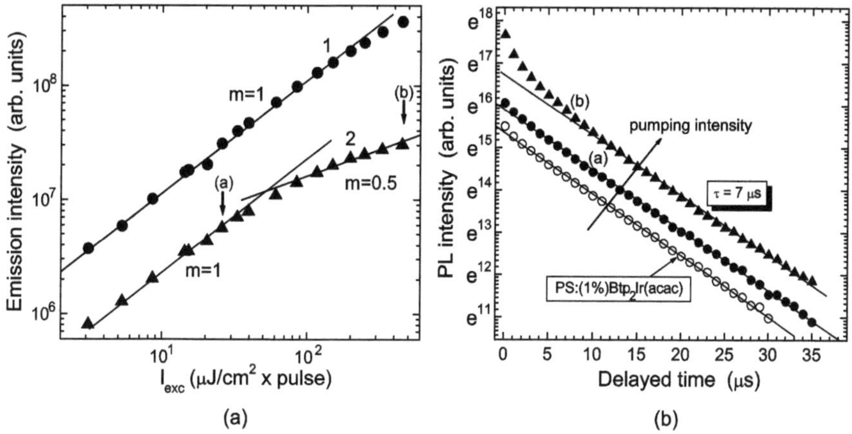

Fig. 6.9 (**a**) Dependence of CNPPP host fluorescence (curve 1) and $Btp_2Ir(acac)$ guest phosphorescence (curve 2) upon excitation intensity at 80 K in a CNPPP:(1 wt%)$Btp_2Ir(acac)$ film. The *arrows* depict the low- and high- laser power intensities at which the Ph decay kinetics presented in Fig. 6.9(b) have been measured. (**b**) Time decay kinetics of Ph emission measured in a CNPPP:(1 wt%)$Btp_2Ir(acac)$ film at low- and high- laser excitation intensity at 80 K (curve (a) and (b), respectively). The gate width was 0.1 μs. The Ph decay kinetics at 80 K of 1 wt% $Btp_2Ir(acac)$ doped into a polystyrene film are given by *open circles* (this curve is arbitrary shifted on the ordinate scale). (© Elsevier)

cay kinetics in this time domain do not follow an exponential decay law. Upon lowering the triplet concentration at $t_{del} > 10$ μs the kinetics become mono-exponential again. The onset of the T-T annihilation process has been straightforwardly observed in the Ph intensity dependence measurements of Fig. 6.9(a). It should be noted that no deviation from the mono-exponential Ph decay with $\tau = 7$ μs was observed when $Btp_2Ir(acac)$ was doped into a PS film, irrespective of the laser excitation intensity level (Fig. 6.9(b), open cycles).

The excitation intensity effect on the Ph yield was studied recently for PtOEP doped in MeLPPP and the Ph yield was found to decrease at high intensity of the incident light [307]. The observed saturation effect of the dopant phosphorescence at intensities exceeding ~10 μJ/(cm^2 pulse) was virtually the same regardless whether PtOEP was doped into MeLPPP or into an optically neutral PS matrix and the phenomenon was explained in terms of bleaching of ground-state PtOEP molecules upon population of their triplet state [307]. Such an explanation is definitely not applicable to the Ph intensity dependence of CNPPP:$Btp_2Ir(acac)$ observed in the present study as the dominance of bleaching is at variance with (i) the observation of the obvious deviation of the Ph decay kinetics from a single exponent at high excitation intensities (Fig. 6.9(b)), which, on the other hand, nicely agrees with the T-T annihilation mechanism; (ii) no decrease of Ph yield of $Btp_2Ir(acac)$ was found when it was embedded in a neutral PS matrix and in frozen diluted solution up to the pumping intensity of ~500 μJ/(cm^2 pulse); (iii) the crossover from a linear to square-root increase of the Ph intensity dependence in the CNPPP:$Btp_2Ir(acac)$ host-guest system tends to shift to lower pumping intensities with increasing temperature, thus pointing to the importance of triplet exciton diffusion. Hence, the reduction of the Ph yield in the CNPPP:$Btp_2Ir(acac)$ system at high excitation intensity is most probably governed by T-T annihilation, which predominates over the possible bleaching effect of the $Btp_2Ir(acac)$ ground-state. The latter effect in the present system might certainly be possible at larger excitation intensity.

The question arises which triplets, of the host or of the guest, are responsible for the T-T annihilation quenching in the CNPPP:$Btp_2Ir(acac)$ system. The results obtained on $Btp_2Ir(acac)$ doped into a PS or a CNPPP host matrix suggest that at 1 wt% doping the guest triplet excitations must be confined by both polymer hosts. At such low concentration, guest triplet migration is not possible, in other words, guest triplets are immobile. Therefore, T_{guest}-T_{guest} annihilation is impossible. On the other hand, all CNPPP host triplet excitons are eventually transferred to the phosphorescent guest as proven by the complete quenching of the intrinsic CNPPP phosphorescence (see further down, Fig. 6.10). This precludes T_{host}-T_{host} annihilation at not very small dopant concentration. Therefore, we conclude that the observed T-T annihilation proceeds via a T_{guest}-T_{host} annihilation reaction, i.e., when a mobile host triplet encounters a localized guest triplet excitation. This picture is in agreement with the absence of T-T annihilation quenching for $Btp_2Ir(acac)$ doped in a PS host matrix at the same concentration level, since excited host triplets are missing in this case. It should be mentioned that bimolecular hetero-annihilation between triplet excitons of a polymer and triplet excitations of dopants has already been reported before for MeLPPP:PtOEP host-guest systems [310].

Figure 6.10 presents time-resolved PL spectra of a CNPPP film doped with 0.05 wt% $Btp_2Ir(acac)$ recorded at 80 K over a very wide range of delay times after the laser pulse excitation (from 0.1 μs to 800 μs) and employing a large width of 1 ms of the gate window (so-called integration mode). As can be seen, both the DF of the host and the Ph of the dopant contribute to the delayed emission spectra. However, no intrinsic Ph of the CNPPP host at 542 nm (Fig. 6.1) could be observed in the CNPPP:$Btp_2Ir(acac)$ host-guest system, implying that the host triplet excitations are efficiently quenched by the lower-energy $Btp_2Ir(acac)$ triplet state via Dexter mechanism. With increasing $Btp_2Ir(acac)$ concentration a reduction of the delayed fluorescence intensity of the polymer host has been observed.

The time dependence of the DF and Ph intensity in the CNPPP:$Btp_2Ir(acac)$ host-guest system is shown in the inset of Fig. 6.10. A remarkable observation is that the Ph of $Btp_2Ir(acac)$ in a CNPPP matrix can be measured even at delay times that are orders of magnitude larger than the intrinsic Ph lifetime of $Btp_2Ir(acac)$ (~7 μs). As can be seen, after the initial decay of the relatively short-lived guest phosphorescence featuring a lifetime of 7 μs at 80 K, the Ph decay kinetics show a very long tail lasting up to milliseconds. This observation of "delayed phosphorescence" may have two plausible origins: (i) delayed arrival of intrinsically long-lived host triplets to the $Btp_2Ir(acac)$ guest molecules, or (ii) delayed production of guest triplets by delayed recombination of (trapped) charge carriers on $Btp_2Ir(acac)$ guest molecules.

Besides delayed guest Ph, intensive delayed fluorescence (DF) of the CNPPP host has also been observed in the delayed emission spectra of CNPPP:(0.05 wt%)$Btp_2Ir(acac)$ films (Fig. 6.10). The DF in these films (as well as in pristine

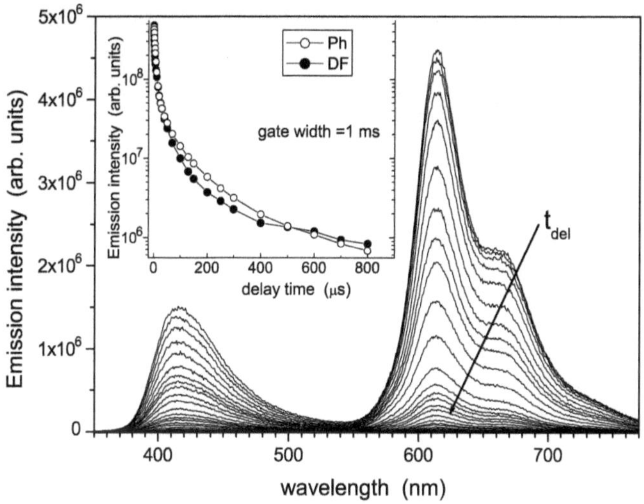

Fig. 6.10 Time-resolved PL spectra of a CNPPP film doped with 0.05 wt% $Btp_2Ir(acac)$ at 80 K. The delay time t_{del} varies from 0.1 μs to 800 μs at a constant gate width of 1 ms. The *inset* shows the time dependence of the DF and Ph intensity in the CNPPP:$Btp_2Ir(acac)$ host-guest system. (© Elsevier)

CNPPP films) arises most probably due to T_{host}-T_{host} annihilation as (i) the intensity dependence of the DF varies almost quadratically with the excitation laser intensity (not shown here), implying a bimolecular mechanism of population of the singlet level, (ii) Ph of the guest is a linear function of the pumping intensity, and (iii) the CNPPP film shows intrinsic long-lived triplet excitations detected even in the millisecond time domain (Fig. 6.1). These factors seem to favor the mechanism of delayed arrival of host triplets as the main cause for delayed phosphorescence. However, at higher $Btp_2Ir(acac)$ concentration the DF emission of the host decreases dramatically. This is in good agreement with the mechanism of T_{host}-T_{host} annihilation as source of the DF. Indeed, as guest molecules readily accept triplets from the host, it is expected that the apparent host triplet lifetime and their population strongly decreases with increasing guest concentration. Consequently, the intensity of delayed T_{host}-T_{host} annihilation, and therefore DF, is reduced at higher doping concentrations. On the other hand, increasing the $Btp_2Ir(acac)$ dopant concentration has almost no effect on the delayed phosphorescence in CNPPP:$Btp_2Ir(acac)$ films. The decay kinetics of the long-lived component (delayed phosphorescence) were found to be virtually the same for different $Btp_2Ir(acac)$ dopant concentrations ranging from 0.05 to 5 wt%. For the mechanism based on delayed arrival of host triplet excitations one should contrarily expect that the decay of the above slow component (delayed phosphorescence) becomes much faster with increasing dopant concentration because quenching of host triplets by triplet-emitter dopants is considerably enhanced in such case. Therefore, we are forced to conclude that the delayed production of guest triplets occurs mainly via delayed recombination of charge carriers on $Btp_2Ir(acac)$. Such an efficient delayed charge-carrier recombination could be ensured by trapped charge carriers and is most probably the dominant mechanism for the delayed $Btp_2Ir(acac)$ phosphorescence. To elucidate the origin of charge trapping in the CNPPP:$Btp_2Ir(acac)$ host-guest system thermally stimulated luminescence (TSL) experiments have been performed.

6.1.2 Thermally Stimulated Luminescence in CNPPP Doped with $Btp_2Ir(acac)$

Figure 6.11 presents the TSL glow curve of a CNPPP film doped with $Btp_2Ir(acac)$ (solid curve). As can be seen, TSL of CNPPP:$Btp_2Ir(acac)$ contains two major peaks, namely a low-temperature peak near 50 K and a high-temperature one with a maximum at 220 K. TSL of a pristine CNPPP film is shown for comparison (dashed curve) and is dominated by the low-temperature TSL peak at 50 K while another weaker peak emerges at ~150 K. It should be mentioned that a similar low-temperature peak centered at 45–50 K has been commonly observed in many conjugated polymers such as MeLPPP [242], polyfluorenes [309], PPV derivatives [243], etc. and it arises due to charge carrier release from the tail states of the intrinsic density-of-state (DOS) distribution of the polymer. The weaker TSL feature at 150 K in pristine CNPPP film (dashed curve) is quite similar to the one observed in polyfluorene polymers [309] and its relative intensity is rather sensitive to film preparation conditions. This weak peak is due to moderately deep inherent charge traps in the CNPPP film, which could be of structural origin in this material. It might be related to some sort of aggregates of polymer chains as it was found before for substituted polyfluorenes [309] and PPV derivatives [243]. As evidenced from Fig. 6.11, doping of the CNPPP host with $Btp_2Ir(acac)$ results in the appearance of a new high-temperature TSL peak at 220 K, suggesting that $Btp_2Ir(acac)$ molecules create rather deep traps in the polymer matrix. The activation energy of this deep trap is about 0.6 eV, as measured by the fractional TSL technique, and agrees well with the difference between the HOMO level of CNPPP and that of $Btp_2Ir(acac)$ which are, respectively, 5.7 [239] and 5.0 eV [240]. Hence, TSL trapping spectroscopy provides a direct evidence that $Btp_2Ir(acac)$ phosphorescent complexes create relatively deep hole traps in the CNPPP matrix. Electron trapping by $Btp_2Ir(acac)$ is not possible because the LUMO of the dopant (2.2 eV [240]) is higher than the one of the polymer matrix (2.3 eV [239]).

The fact that $Btp_2Ir(acac)$ complexes act as hole traps in CNPPP, support our argumentation that the delayed production of guest triplets occurs mainly via delayed recombination of charge carriers and may explain the observation of delayed phosphorescence (Fig. 6.10). When a hole is trapped on $Btp_2Ir(acac)$ molecules and an electron is localized on some shallower intrinsic traps of the CNPPP polymer, trapped geminate electron-hole pairs can be readily generated. Such trapped geminate pairs are typically more stable against recombination compared to non-trapped electron-hole pairs and depending on mutual separation between opposite sign charge carriers they can survive for milliseconds or even seconds [242, 330].

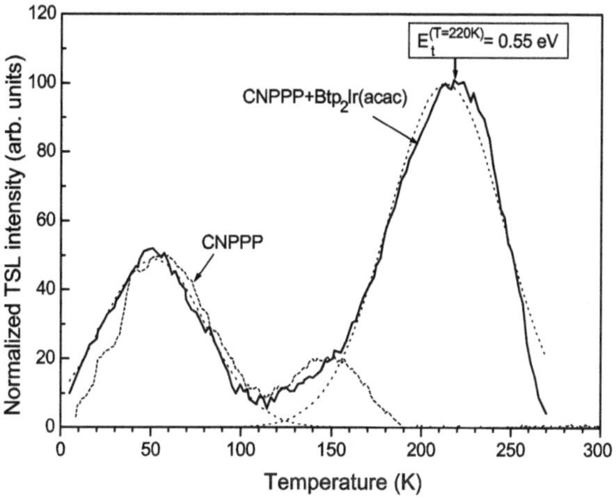

Fig. 6.11 The TSL glow curve of a CNPPP film doped with 5 wt% $Btp_2Ir(acac)$ (*full line*) and pristine CNPPP film (*dashed line*) after excitation with 365 nm light at 4.2 K. (© Elsevier)

Since recombination occurs directly on (or in the vicinity of) the dopant sites initiated by a trapped hole, it is natural that it produces guest phosphorescence, which follows the charge recombination kinetics. In addition, charge carrier trapping by $Btp_2Ir(acac)$ molecules can in an excellent way explain the more efficient quenching of the host fluorescence component in the electroluminescence spectra of CNPPP:$Btp_2Ir(acac)$ OLEDs compared to the situation in photoluminescent spectra (Fig. 6.8(a) and (b)). Charge carrier trapping has often been used before to explain this phenomenon [328, 329]; however, in the present study direct evidence that the phosphorescent metal-organic complexes do act as charge traps was presented by means of TSL spectroscopy.

6.2 Optical Pumping of Triplet-Emitters

An important loss mechanism specific for electrically operating light-emitting devices under severe bias is singlet-triplet annihilation [50]. The longer lifetime of triplets compared to singlets and the fact that under electrical excitation much more triplets than singlets are generated, contribute to a high steady-state triplet density in the device. At high brightness conditions the triplet concentration can become high enough to quench singlets severely [50, 212, 215]. Using triplets for light-emission is an appealing solution to this problem.

In this section, the route of using triplet-emitters as light-emitting species in the active layer of an organic laser is investigated. The advantages are the smaller linewidth of the spontaneous emission spectrum of phosphorescent compounds and the larger internal quantum efficiency [34]. As a result of both, the required current for lasing could be one order of magnitude smaller using triplet emitters. Lasing either by optical or electrical pumping of triplet-emitters, however, has not been demonstrated so far. Here, three different phosphorescent compounds are examined for signs of stimulated emission.

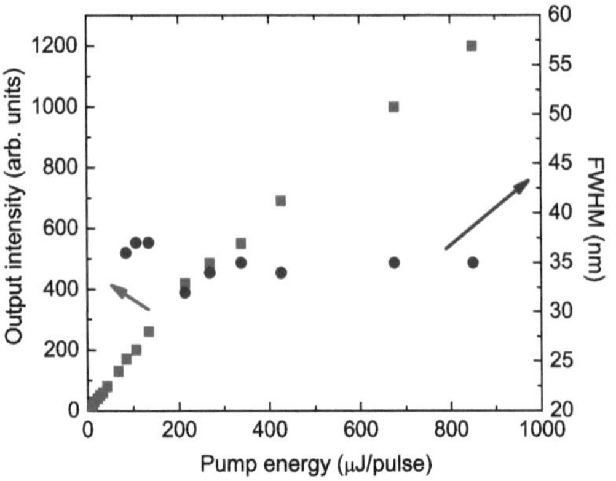

Fig. 6.12 Emission intensity and associated FWHM of a 600-nm thick PMMA:(2 wt%)$Btp_2Ir(acac)$ film measured at various pumping intensities. (Measurements performed by RWTH Aachen)

6.2.1 Attempts to Observe ASE Using $Btp_2Ir(acac)$

There are different options to check whether ASE can be observed from the commercially available phosphorescent metal-organic complex $Btp_2Ir(acac)$. The ideal situation is to use an inert film doped with a small concentration of the triplet-emitter. Because absorption is expected to be very low due to the low dopant level and the small extinction coefficient of $Btp_2Ir(acac)$, a relatively thick film should be used. Using a host-guest system, e.g. CNPPP:$Btp_2Ir(acac)$, could be another option. This has been carefully investigated in Sect. 6.1. Alternatively, ASE might be examined in diluted solution. All three possibilities were investigated.

First, $Btp_2Ir(acac)$ was dispersed in PMMA[3] and PS, which are considered as optically inert matrices. Both matrices were extensively studied for the dispersion of laser dyes, given their very good optical and thermal properties [125, 331]. A relatively thick film of 600 nm comprising 2 wt% $Btp_2Ir(acac)$ was prepared. Figure 6.12 shows the emission intensity of this film and the associated FWHM measured at various pumping intensities. Even at very high pumping intensities (well above 10 mJ/pulse) there could not be any increase of the slope efficiency observed. In fact, there was even a slight decrease of the emission intensity above 200 μJ/pulse, which might be attributed to sample degradation. In addition, no spectral narrowing was observed as indicated by a constant FWHM of the measured emission spectra. Similar results were obtained when a CNPPP:$Btp_2Ir(acac)$ film and a diluted solution of $Btp_2Ir(acac)$ molecules were optically pumped, indicating that it was impossible to obtain stimulated emission using the phosphorescent compound $Btp_2Ir(acac)$.

In order to identify the mechanism responsible for the lack of ASE, time-resolved pump-probe measurements were performed on a 4 μm thick PMMA:$Btp_2Ir(acac)$ film deposited on a quartz substrate. The sample was pumped at a wavelength of 400 nm, while the wavelength of the probe beam was set at the PL emission maximum $\lambda = 614$ nm. Figure 6.13 depicts the result of this experiment at an excitation

[3]Poly(methyl metacrylate).

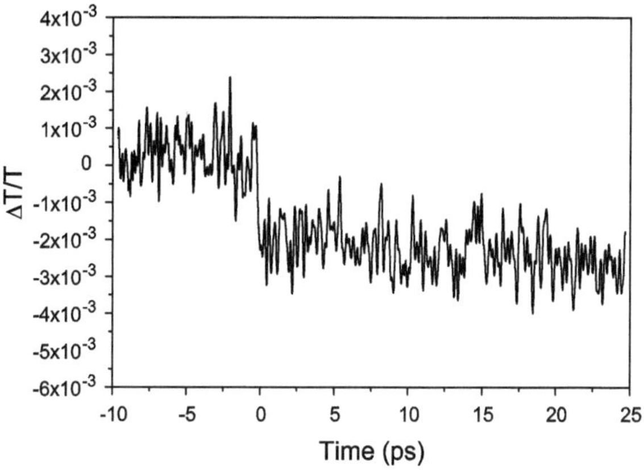

Fig. 6.13 Time-resolved pump-probe experiment carried out on a 4 μm-thick PMMA sample doped with 2 wt% $Btp_2Ir(acac)$. Sample excitation occurred at 400 nm using a laser fluence of 195 μJ/cm^2; the wavelength of the probe beam was set to 614 nm. (Measurements performed by RWTH Aachen)

laser fluence of 195 μJ/cm^2. As can be seen, a clear reduction in the transmission upon photo-excitation is measured. This observation indicates that $Btp_2Ir(acac)$ exhibits excited state absorption, which competes with radiative emission, and therefore prevents ASE. Due to fast intersystem crossing of singlets to triplet states, the excited state absorption of $Btp_2Ir(acac)$ is dominated by triplet-triplet absorption. These findings are in agreement with measurements reported by Adachi *et al.*, who demonstrated that several commonly-used Ir-complexes have large excited state absorptions [332] at the spectral region where light emission occurs. A cross-section for triplet-triplet absorption σ_{TT} of $3 \cdot 10^{-18}$ has been reported for $Btp_2Ir(acac)$, which is much larger than its $\sigma_{se} \simeq 7.9 \cdot 10^{-20}$ precluding any lasing action [332]. Apart from photoinduced absorption losses, also non-radiative losses such as quenching by oxygen and T-T annihilation due to the presence of movable host triplets (Sect. 6.1.1) may additionally deplete the triplet reservoir.

6.2.2 Attempts to Observe ASE Using F5Ph and GDP16b

The large σ_{TT} at the expected lasing wavelength of commercially available triplet-emitters, such as $Btp_2Ir(acac)$, might be problematic for achieving lasing action. In order to stand chance of obtaining ASE from triplet-emitters, the effective radiative lifetime has to be decreased (to have a larger cross-section for stimulated emission σ_{se}) or/and the spectral region where triplet-triplet absorption occurs has to be shifted with respect to the light emission spectrum. The metal-organic complexes F5Ph and GDP16b partially fulfill these requirements and were therefore optically pumped to check if ASE could be observed.

F5Ph was chosen because of its small radiative lifetime (∼3 μs) compared to other triplet-emitters. However, its transient absorption spectrum overlaps with the spectral region where light-emission occurs. To check whether it is possible to observe ASE from F5Ph, two different experiments were performed. First, a thick (∼1 μm) film of PMMA doped with 6 wt% F5Ph was measured. The results are

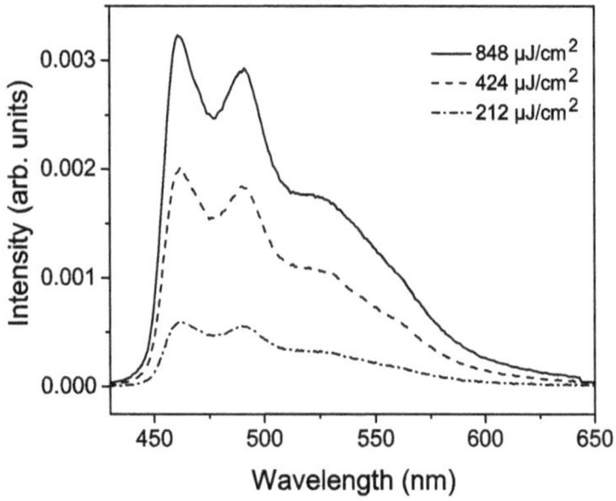

Fig. 6.14 Emission spectra of F5Ph for different laser fluences at 400 nm excitation wavelength. (Measurements performed by RWTH Aachen)

shown in Fig. 6.14. As can be seen, no spectral narrowing could be observed. Additionally, a diluted solution of 1 wt% F5Ph in de-aerated toluene has been measured at room temperature. Triplet-triplet annihilation and quenching of triplets by oxygen could definitely not be an issue in this case, however, no ASE was observed. We attribute these results to the high triplet-triplet absorption in the spectral region of interest.

In contrast to F5Ph, the metal-organic compound GDP16b exhibits triplet-triplet absorption which is reduced at the light-emission spectral range. The photoluminescence spectrum as well as the transient absorption spectrum of this compound are depicted in Fig. 6.15(a). Emission occurs in the range of 600–800 nm, which corresponds to a region of low transient absorption. Optical pumping of a thick PMMA film doped with 6 wt% GDP16b, however, failed to show spectral narrowing (Fig. 6.15(b)). This suggests that only reducing triplet-triplet absorption might not be enough to achieve lasing in triplet-emitters. Another possible explanation for the absence of ASE in GDP16b doped PMMA films, could be the presence of oxygen or the relatively high GDP16b dopant concentration.

6.3 Summary and Conclusions

In this chapter, triplet-emitter doped organic materials were studied. Triplet dynamics and charge carrier trapping in CNPPP and PS films incorporating the red-emitting metal-organic complex $Btp_2Ir(acac)$ were investigated. In addition, the intrinsic phosphorescence (Ph) of CNPPP was characterized. The results show that the decrease in the Ph efficiency in CNPPP:$Btp_2Ir(acac)$ systems at increasing excitation intensity is dominated by mutual host-guest triplet-triplet annihilation. Spectral diffusion effects were observed at 10 wt% of $Btp_2Ir(acac)$ dispersed in a polymer matrix, which is a direct proof of triplet migration though the manifold triplet-emitter sites at such dopant concentration level. This observation suggests straightforwardly that the concentration quenching observed in PS:$Btp_2Ir(acac)$ films arises due to triplet exciton migration at increased guest molecule concentration and thus

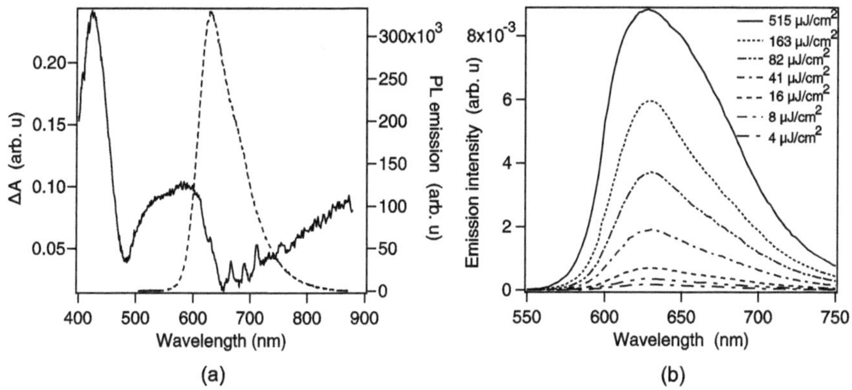

Fig. 6.15 (**a**) Emission spectrum (*dotted line*) and transient absorption spectrum measured at 400 ns delay time (*full line*) of GDP16b, and (**b**) emission spectra of a 6 wt% GDP16b doped PMMA film, measured at various laser fluences. (Measurements performed by RWTH Aachen)

their ability to encounter non-radiative oxygen impurities. A "delayed phosphorescence" was found in CNPPP:$Btp_2Ir(acac)$ films, which emerged at a time delay orders of magnitude larger than the natural lifetime of $Btp_2Ir(acac)$ triplet excitations. Delayed production of guest triplets by recombination of trapped charge carriers on $Btp_2Ir(acac)$ guest molecules was identified as the dominant mechanism for the observed delayed phosphorescence. The latter was supported by TSL measurements, which provided direct evidence that $Btp_2Ir(acac)$ creates hole traps in the conjugated polymer CNPPP. Good agreement between the experimentally obtained trap depth and the one predicted from the difference in HOMO levels was found.

In the second section of this chapter, the use of triplet-emitters as light-emitting species in the active layer of organic laser was investigated. Photoluminescence and time-resolved spectroscopic experiments on $Btp_2Ir(acac)$ showed the absence of amplified spontaneous emission, even when the samples were excited at very high pumping intensities. The reason for this behavior was attributed to an effective excited state absorption, which competes with radiative emission, and therefore prevents ASE. Additionally, two other metal-organic compounds, F5Ph and GDP16b, which exhibit, respectively, a relatively short radiative lifetime and a shifted transient absorption spectrum with respect to the emission spectrum were examined. However, also by using these compounds no spectral narrowing could be observed.

Chapter 7
Value of OLEDs with Field-Effect Electron Transport for Lasing Applications

Organic light-emitting diodes (OLEDs) [84, 96, 215] and organic light-emitting field-effect transistors (LEOFETs) [49, 65, 75, 221, 222] are two device architectures that have been proposed as potential device configuration for future electrically pumped lasers. In this work, OLEDs with field-effect electron transport were added to the list of existing electroluminescent devices. The aim of the present chapter is to investigate the prospects of realizing an electrically pumped laser using these OLEDs with field-effect electron transport. First, the potential of the structure as laser device configuration is analyzed. The advantages as well as the existing limitations are highlighted. Next, amplified spontaneous emission measurements and optically pumped lasing experiments are performed. Stimulated emission in the host-guest system Alq_3:DCM_2 and in organic layer stacks including hole- and electron-transporting layers used in OLEDs with field-effect-assisted electron transport are studied. In the end, the different possibilities to incorporate an optical feedback mechanism in the device are discussed.

7.1 Potential of OLEDs with Field-Effect Electron Transport as Laser Device Configuration

OLEDs with field-effect electron transport exhibit several characteristics that are promising with respect to the realization of an electrically pumped organic laser. The high current density that can be achieved without significant reduction of the external quantum efficiency and the reduced optical absorption losses due to the remoteness of the metallic cathode, are the most important ones and are discussed in detail in Sect. 7.1.1 and Sect. 7.1.2, respectively.

Apart from minimizing the optical absorption losses and allowing high current densities, OLEDs with field-effect electron transport may also be able to reduce the triplet population in the device. In Chap. 5 (Sect. 5.1) it has been demonstrated that pulses as low as 1 μs could be applied to the device without affecting the light intensity. Since this pulse width is much shorter than the typical triplet lifetime of

S. Schols, *Device Architecture and Materials for Organic Light-Emitting Devices*,
DOI 10.1007/978-94-007-1608-7_7, © Springer Science+Business Media B.V. 2011

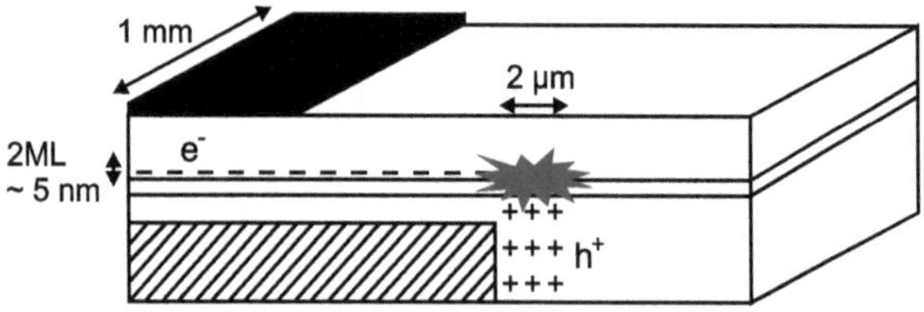

Fig. 7.1 Illustration of the areas taken into account to calculate the electron and hole current density

fluorescent materials, pulsed excitation of the device may reduce triplet accumulation. In addition, since the electrons are transported by field-effect, high mobilities are achieved, reducing the number of charge carriers needed to generate population inversion in the device. The combination of all these properties makes OLEDs with field-effect electron transport interesting for waveguide devices and future electrically pumped lasers.

7.1.1 High Current Densities

In Chap. 3 and Chap. 4 it has been demonstrated that very high current densities can be achieved in OLEDs with field-effect electron transport without significant reduction of the external quantum efficiency. Using the electron-transporting material PTCDI-$C_{13}H_{27}$ a maximum electron and hole current density of, respectively, 5400 A/cm^2 and 13 A/cm^2 could be measured. The large difference between both values is due to the fact that for the calculation of the electron current density the thickness of the accumulation layer, which is about 5 nm [45, 256], was used, whereas for the hole current density the lateral area of the recombination zone (~2 µm, Sect. 3.3) was taken into account. This is illustrated in Fig. 7.1.

The feasibility of electrically pumped lasing using OLEDs with field-effect electron transport can be investigated by calculating the exciton density per unit area in the recombination zone σ_{RZ}. This exciton density can be estimated using [69]:

$$\sigma_{RZ} = \frac{J}{q} \cdot r_{st} \phi_{PL} \tau, \tag{7.1}$$

where J is the current density, q is the elementary charge, r_{st} is the singlet/triplet ratio of excitons, ϕ_{PL} is the photoluminescence efficiency and τ is the exciton lifetime. For lasing it is the hole current density of 13 A/cm^2 that is most relevant because this takes into account the amount of excitons generated over the entire width of the recombination zone. The photoluminescence efficiency of DCM_2 was measured to be about 40% and is in agreement with previous reports [41]. The exciton lifetime of DCM_2 is 1 ns [333] and for fluorescent devices only a quarter of the injected charge carriers forms singlet excitons [36]. Using these values an exciton density of $8 \cdot 10^9$ cm^{-2} could be calculated. This value is lower than the typical values of exciton densities needed to reach the threshold in optically pumped lasers (~10^{11}–10^{12} cm^{-2} [69, 184, 221]), indicating that the achievable current density in OLEDs with field-effect electron transport is still insufficient to allow electrically pumped lasing.

The maximum achieved exciton density per unit area of our device concept is, however, still higher compared to other promising device architectures. Naber *et al.* reported a σ_{RZ} value of $1 \cdot 10^9$ cm^{-2} for polymer light-emitting field-effect transistors using a combination of a low-k and a high-k material as gate dielectric [69]. For the tetracene-based single crystal LEOFETs of Takenobu *et al.* they calculated $\sigma_{RZ} = 5 \cdot 10^6$ cm^{-2} [69]. Performing the same calculation for the LEOFET reported by Namdas *et al.* [63], using I (50 µA), W (1000 µm), W_{RZ} (4 µm), ϕ_{PL} (53% [221]) and τ (5 ns [221]) gives $\sigma_{RZ} = 5 \cdot 10^9$ cm^{-2}. The slightly higher σ_{RZ} achievable in our devices is mainly attributed to the higher current densities that can be obtained and might be further improved by reducing the distance to the insulator edge, by using pulsed operation to higher voltages, by employing high-k dielectrics with low leakage current, by optimizing the device geometry and by using other light-emitting materials with a higher photoluminescence efficiency and/or a larger exciton lifetime. Note that the incorporation of triplet-emitters might also enhance σ_{RZ} since in this case $r_{st} = 100\%$, although other loss mechanisms such as triplet-triplet absorption and triplet-triplet annihilation may become dominant in that case (Sect. 6.2).

7.1.2 Reduced Absorption Losses

A main challenge to overcome with respect to the realization of electrically pumped lasing are the additional losses due to absorption of photons at the electrical contacts. In thin waveguide structures, where the resonator is positioned in the plane of the film, this is particularly important because of the long interaction between the light and the electrodes. OLEDs with field-effect electron transport may significantly reduce these detrimental waveguide losses, because the metallic cathode is displaced from the light-emitting zone.

Numerical simulations using the free full-vectorial Maxwell solver CAMFR[1] allowed us to compare the waveguide losses in OLEDs with field-effect electron transport to those in standard OLEDs. In CAMFR the materials are characterized by complex refractive indices. For the Al cathode and the ITO anode a value of, respectively, 1.040-6.189j and 1.823-0.003j was used. The imaginary part of the refractive index of the organic layer was assumed negligible. An organic layer with a thickness of 200 nm was taken for both device architectures. The results of the simulations are summarized in Table 7.1. Clearly, the OLED with field-effect electron transport shows a significant reduction of the optical absorption losses compared to conventional OLED structures, indicating that the absence of Al in the vicinity of the recombination zone indeed substantially decreases the waveguide losses. This reduction becomes more important at decreasing ITO thickness. When the devices comprise an anode of only 20 nm ITO, the waveguide losses are reduced by a factor ten.

---

[1] CAvity Modelling FRamework (developed at INTEC, Ghent University, Belgium).

Table 7.1 Waveguide losses in conventional OLEDs and OLEDs with field-effect electron transport for various thicknesses of the ITO anode

	OLED	OLED with field-effect transport
ITO = 100 nm	256 cm^{-1}	104 cm^{-1}
ITO = 20 nm	222 cm^{-1}	21 cm^{-1}

The influence of the contact materials on the properties of the organic waveguide can also be experimentally analyzed by a combination of ASE and waveguide propagation loss measurements. For this purpose a series of multi-layer slab waveguide samples based on 150-nm thick MeLPPP films were prepared. MeLPPP was chosen because of its excellent film-forming properties, its high photoluminescence quantum efficiency in solid-state (~30%) and its large gain coefficient [157]. For each of the fabricated samples, the emission spectra for various excitation fluences were recorded. Plotting the emitted intensity at the peak wavelength ($\lambda = 495$ nm) as a function of the excitation intensity allowed to determine the ASE thresholds, as displayed in Fig. 7.2.

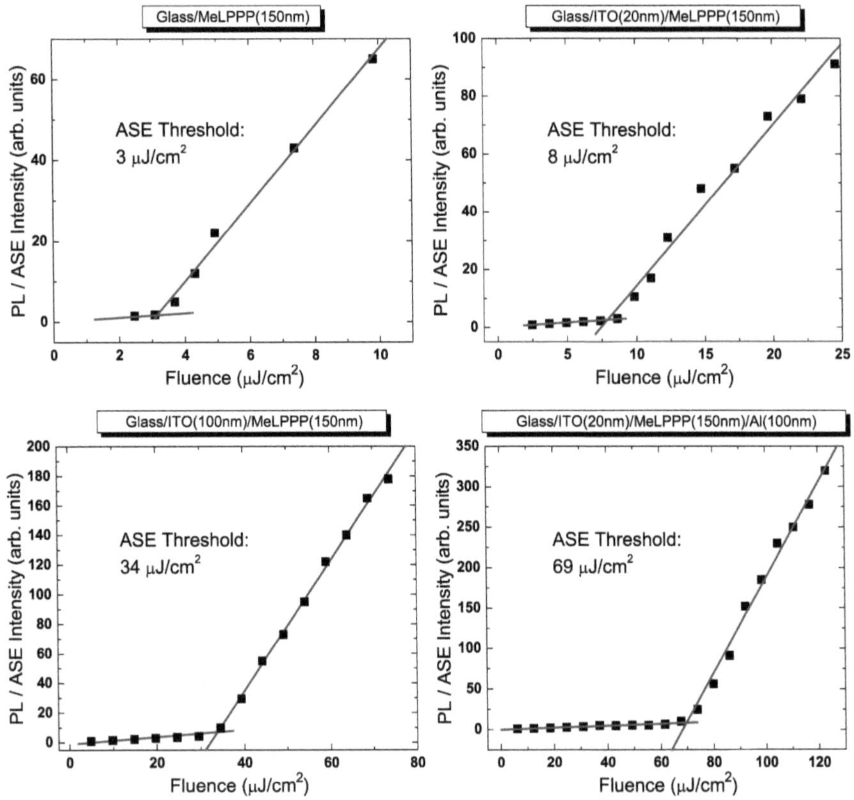

Fig. 7.2 ASE threshold measurements of different MeLPPP waveguide samples. (Measurements performed by RWTH Aachen)

The ASE threshold is determined as the excitation intensity at which the output signal slope efficiency abruptly increases. Clearly, the lowest ASE threshold of 3 μJ/cm^2 is achieved when MeLPPP is deposited on a bare glass substrate. A 20 and 100 nm thick ITO layer sandwiched between the organic film and the substrate results in an increase of the ASE threshold to, respectively, 8 and 34 μJ/cm^2. An additional Al layer on top of the organic waveguide further increases the ASE threshold to 69 μJ/cm^2. Note, that all samples were optically pumped from the back side, i.e. through the glass substrate. Therefore, strong absorption of the pump light by the metallic Al layer can be neglected when comparing the different samples.

One can expect that the observed increase in ASE threshold is based on a reduction of the optical confinement factor in the gain medium (reduced gain) and/or the extension of the waveguide modes into the absorbing contact layers (increased waveguide losses). Loss measurements, carried out by the experimental technique introduced in Chap. 2 (Sect. 2.2.5), allowed to determine the loss coefficient of the different waveguide structures. The results of these loss measurements, which actually provide information about the propagation of light in the unpumped region of the waveguides, are shown in Fig. 7.3. The graphs depict the edge emitted light intensity at $\lambda_{ASE} = 495$ nm as a function of the distance between the waveguide edge and the excitation stripe. For each sample the laser fluence was chosen above its respective ASE threshold. The data were fitted by an exponential decay function (solid lines) to extract the effective waveguide loss coefficients for the different layer stacks. Waveguide structures comprising absorbing layers adjacent to the MeLPPP organic layer show a significant increase in loss coefficients. The lowest waveguide losses of 5.3 cm^{-1} were observed for MeLPPP on a bare glass substrate. An additional ITO layer sandwiched between the organic film and the substrate increases the waveguide losses to 10.6 and 23.6 cm^{-1} for, respectively, 20 and 100 nm ITO thickness. The presence of an Al layer on top of the organic waveguide further increases the measured waveguide loss to the largest value of 47.6 cm^{-1}.

The first two columns of Table 7.2 summarize the measured ASE thresholds and waveguide losses of the investigated samples. As can be seen, higher ASE thresholds are accompanied by increased waveguide losses. The presence of Al has a large influence on these losses. Hence, OLEDs with field-effect electron transport, which have a metallic top contact displaced with respect to the recombination zone, will exhibit reduced optical absorption losses compared to conventional OLEDs, where light is generated very close to the metallic cathode. The measurements also show that the waveguide losses can be additionally reduced by decreasing the ITO layer thickness, which is in agreement with the simulated results obtained via CAMFR.

For a further evaluation of these results, the intensity profile of the light, which is propagating in the different structures, was simulated using a 2D beam propagation simulation tool (RSOFT BeamPROP). The optical constants of the involved materials were derived from ellipsometry measurements. The simulation results are displayed in Fig. 7.4. In each graph, the black line represents the vertical refractive index profile of the device structure. In lateral direction the sample is assumed to be infinite. Red lines represent the simulated vertical beam intensity profile within the structure. The amount of light guided within the active organic layer is marked by

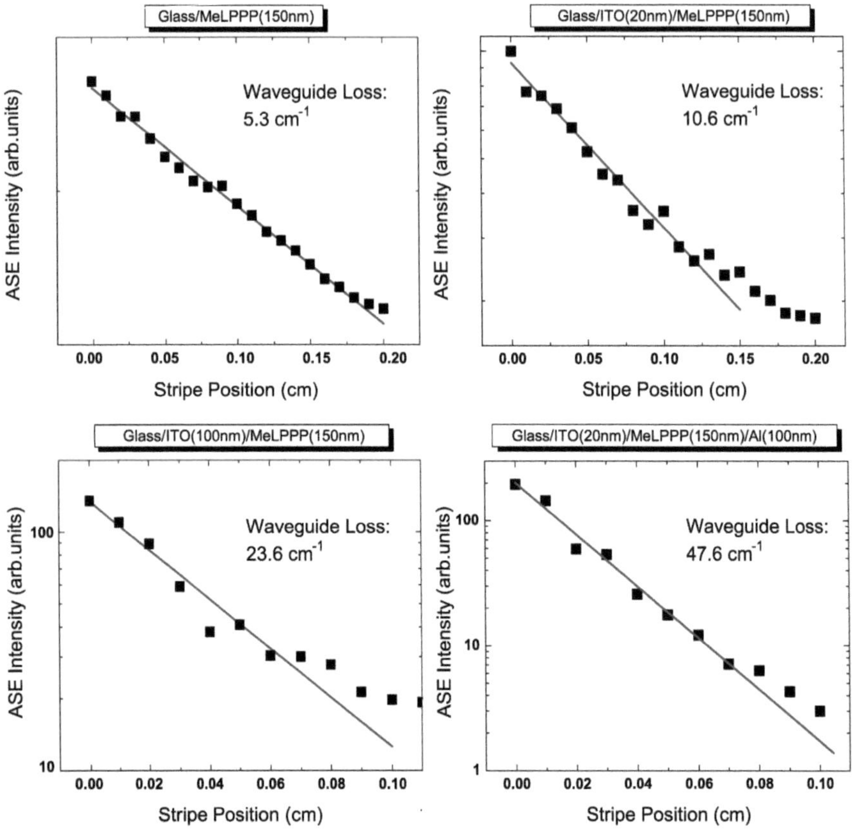

Fig. 7.3 Propagation loss measurements carried out for different MeLPPP waveguide samples. (Measurements performed by RWTH Aachen)

Table 7.2 ASE thresholds, loss coefficients and confinement factor of the investigated slab waveguide samples

sample	ASE threshold	loss coefficient	Γ
MeLPPP/glass	3 μJ/cm^2	5.3 cm^{-1}	53%
MeLPPP/ITO(20 nm)/glass	8 μJ/cm^2	10.6 cm^{-1}	50%
MeLPPP/ITO(100 nm)/glass	34 μJ/cm^2	23.6 cm^{-1}	30%
Al/MeLPPP/ITO(20 nm)/glass	69 μJ/cm^2	47.6 cm^{-1}	37%

the hatched area and the corresponding confinement factor Γ as well as the effective refractive indices n_{eff} of the whole structure are indicated. The confinement factors are summarized in the third column of Table 7.2.

The simulated data can be discussed with respect to their role on either lasing threshold or propagation losses. The losses of a certain structure are mainly de-

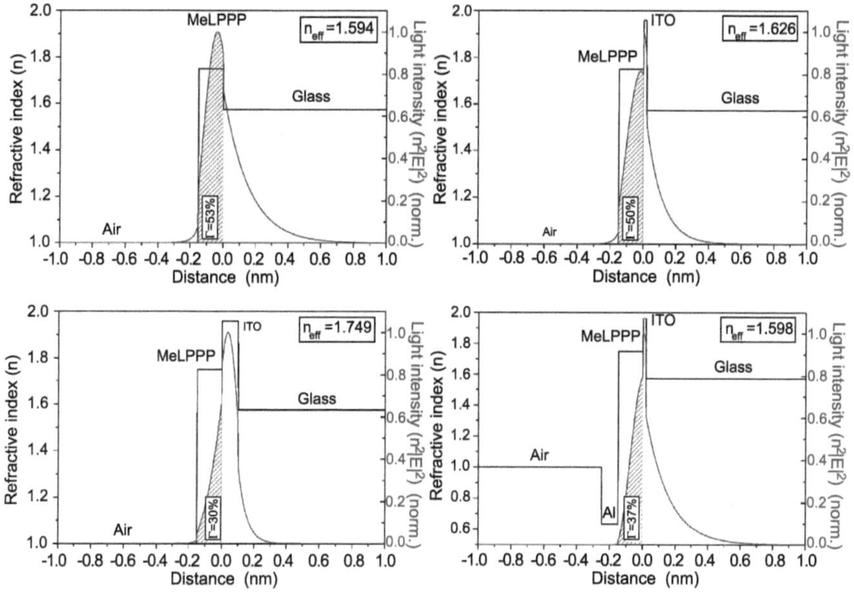

Fig. 7.4 (Color online) Simulated guided mode profiles for different MeLPPP waveguide samples. (Simulations performed by RWTH Aachen)

termined by the percentage of light, which is guided within the lossy ITO and Al layers. For the ASE threshold, the amount of light guided within the active MeLPPP film (confinement factor) has to be taken into account additionally. A comparison between the different samples shows that the presence of 20 nm ITO only slightly decreases the confinement factor compared to the bare MeLPPP film. Consequently, the increase in ASE threshold from 3 μJ/cm^2 to 8 μJ/cm^2 for the former is mainly due to increased waveguide losses in this structure. When increasing the ITO thickness to 100 nm, the confinement factor drops from about 50% to 30%. In addition, the amount of light guided in the ITO layer and thus the accompanied waveguide losses strongly increase. Compared to the waveguide comprising a bare MeLPPP film, the ASE threshold increases by a factor of ten while the measured waveguide losses increase by only a factor of 4–5. This indicates that both the loss of confinement and the severe increase of waveguide losses are responsible for the order-of-magnitude increase in ASE threshold. Finally, the waveguide structure using Al has a confinement factor slightly better than the latter structure and the larger ASE threshold is clearly to be attributed to the waveguide losses provoked by the Al layer.

7.2 Optically Pumped Lasing Experiments

The active light-emitting layer of the OLEDs with field-effect electron transport presented in this work all consisted of an Alq_3 host matrix incorporating 2% of the

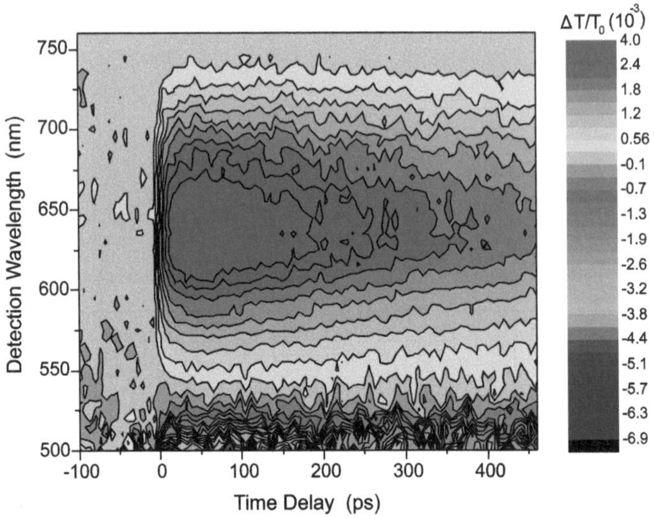

Fig. 7.5 Transient absorption measurement on a 150-nm thick Alq_3:DCM_2 layer, showing a maximum gain near 645 nm. (Measurements performed by RWTH Aachen)

red-emitting dye DCM_2. Alq_3:DCM_2 is a benchmark host-guest system for which low ASE and lasing thresholds have been reported [114, 184]. In this section, stimulated emission of Alq_3:DCM_2 is investigated in single layers as well as in organic layer stacks including hole- and electron-transporting layers.

7.2.1 Stimulated Emission in Alq_3:DCM_2

To verify the wavelength with maximum gain of the host-guest system Alq_3:DCM_2 time-resolved pump-probe measurements have been performed on a 150-nm thick layer of Alq_3 doped with 2% DCM_2, which was deposited on a quartz substrate. For these measurements, the excitation power was set to 110 $\mu J/cm^2$. At this value ASE emission became well observable, but it was low enough to prevent sample degradation over the measurement's duration of 12 hours. The measured time-resolved transient absorption spectrum is presented in Fig. 7.5. Maximum gain is obtained at the spectral region around 640–650 nm. This gain region appears to be red-shifted compared to the photoluminescence spectrum which has its peak between 610–620 nm.

In a complimentary experiment, ASE has been observed in the same wavelength range. This is illustrated in Fig. 7.6, which shows the measured edge-emitted peak intensity of a 150 nm-thick Alq_3:DCM_2 film as a function of the pumping intensity. Above a pumping intensity of 6.6 $\mu J/cm^2$, the slope efficiency of the output signal abruptly increases indicating that spontaneously emitted photons are exponentially amplified by stimulated emission. Since photons whose energy coincide with the spectral position of maximum gain will be amplified more than others, a collapse

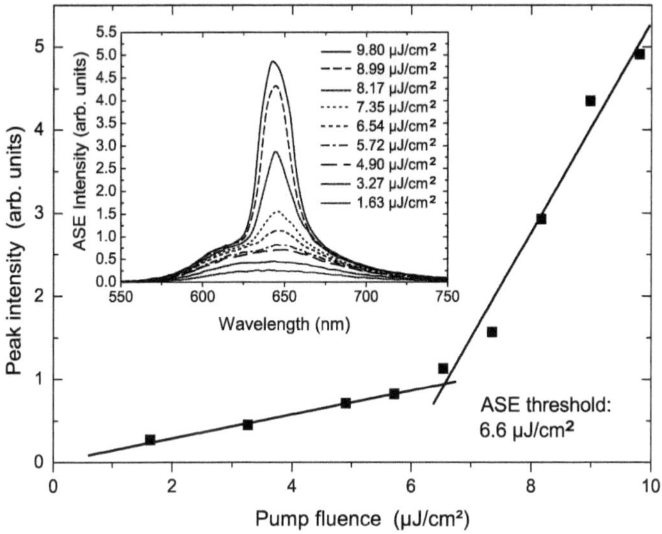

Fig. 7.6 Measurement of the edge emission as a function of the pump power. *Inset*: Evolution of the emission spectrum upon excitation density. Above the ASE threshold narrowing of the spectrum is observed. (Measurements performed by RWTH Aachen)

of the emission spectrum near 645 nm is observed. The evolution of the emission spectrum upon excitation density is shown as inset of Fig. 7.6.

Both measurements indicate that an optical feedback mechanism should be designed to have the resonator wavelength around 640–650 nm. While second-order photonic feedback structures are easier to characterize because of the vertical emission direction, first-order structures have a potentially lower lasing threshold because of the lack of vertical losses [146]. Hence, first-order linear grating structures were fabricated and used to perform optically pumped lasing experiments. Details about grating fabrication are discussed in Chap. 2 (Sect. 2.1.3). From photonic simulations the optimum photonic grating structure for the given technological and optical constraints was obtained. According to these simulations trenches of 50 nm depth were etched into SiO_2 with a period of 210 nm and a duty cycle of 50%. Figure 7.7 shows a SEM image of the cleaved cross section of the fabricated structure after organic layer deposition. Because the waviness of the grating is retained on the surface, an effective average thickness must be taken into account in the photonic simulations. Deposition of 200 nm organic material on top of the grating yields an effective thickness of 225 nm, which should result in the lowest threshold possible and a lasing wavelength between 640–645 nm.

Figure 7.8 shows the emission from 200 nm Alq_3:DCM_2 on the photonic grating structure. Lasing occurred at a wavelength of 639 nm, which is close to the optimum wavelength range between 640–645 nm. A lasing threshold of 110 pJ/pulse ($\sim$4 μJ/cm^2) and a slope efficiency of 0.0017 pJ^{-1} were extracted from the luminescence data. Because the grating was in first order and the lasing mode was guided in-plane, the sample had to be tilted with respect to the light collection lens by ap-

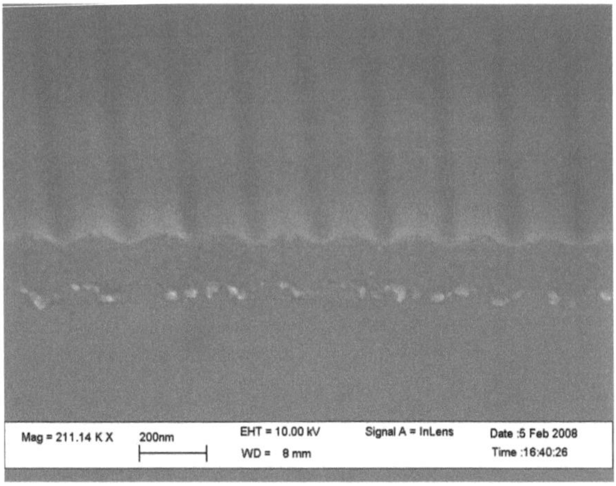

Fig. 7.7 SEM cross-section of the fabricated photonic structure covered with 200 nm Alq_3:DCM_2. (Image received from IBM Research Zurich)

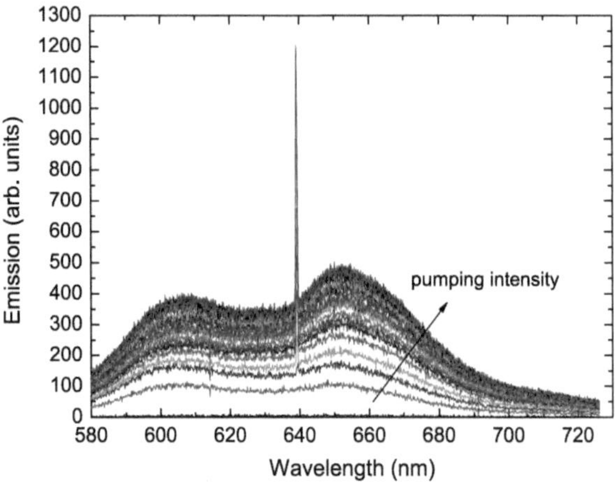

Fig. 7.8 Lasing emission from a first order photonic grating structure covered with 200 nm Alq_3:DCM_2. A lasing peak emerges from the photoluminescence spectrum above the pump power threshold. The measurements were performed at room temperature using a pump wavelength of 400 nm. (Measurements done by IBM Research Zurich)

proximately 30° in order to be able to detect sufficient scattered light intensity. This tilting angle did not have a significant influence on the extracted values, but dramatically improved the signal-to-noise ratio. The presented lasing experiments were performed in close collaboration with IBM Research in Zurich.

7.2.2 Stimulated Emission in Stacks Comprising Alq_3:DCM_2

In OLEDs with field-effect electron transport the active light-emitting region has a thickness of about 20 nm. Such a thin gain layer implies a very low optical confinement factor in the vertical direction ($\Gamma \sim 4\%$), and currently it has not been shown that this can be successfully used to achieve optically pumped lasing in organic films. In order to assess this issue, a series of samples with Alq_3 layers on first order photonic gratings were fabricated in which the region doped with DCM_2

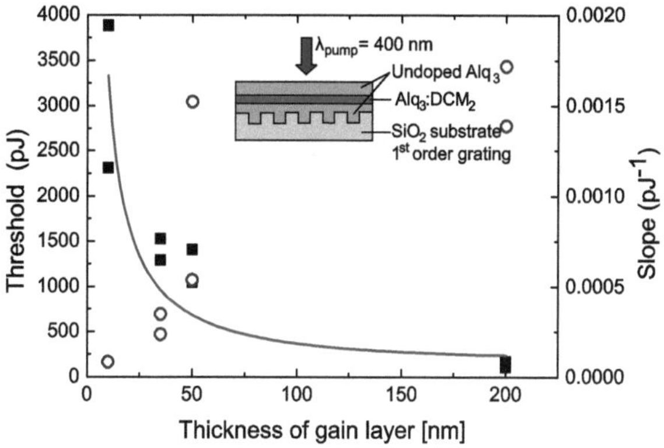

Fig. 7.9 Measured laser threshold (*squares*) and slope efficiency (*open circles*) as a function of thickness of gain layer. The *solid line* is the theoretical laser threshold expected from the calculated confinement factor. *Inset*: arrangement of the organic layer stack. (Measurements performed by IBM Research Zurich)

was systematically varied. The gain regions were always vertically centered in the layer stack in order to achieve good spatial overlap with the intensity maximum of the laser mode. This arrangement is illustrated in the inset of Fig. 7.9. The total thickness of the organic layer stack was 200 nm for all fabricated samples. The lasing behavior of this series of samples was investigated at room temperature using a femtosecond pulse laser having a pump wavelength of 400 nm.

Figure 7.9 displays the laser threshold and the slope efficiency as a function of thickness of the DCM_2-doped part. Lasing could be observed down to gain layer thicknesses of 10 nm. This is an interesting result since lasing for such thin organic gain layers has, to the best of our knowledge, never been demonstrated before. Compared to the device where the whole deposited organic layer shows gain, the laser threshold is, however, about one order of magnitude higher. This was expected from the calculated modal overlap. Accordingly, the slope efficiency of the laser decreases with decreasing gain layer thickness.

Next, the hole-transporting layer PTAA was included in the organic layer stack. 50 nm PTAA was spin-coated on top of a first order grating. Since the refractive indices of PTAA and Alq_3 match closely, the grating geometry could be retained. On top of the PTAA, varying thicknesses of DCM_2-doped Alq_3 were deposited, and the layer stack was completed to a total thickness of 200 nm by undoped Alq_3 as electron-transporting layer.

The luminescence spectra of a sample comprising 50 nm PTAA and 150 nm Alq_3:DCM_2, measured at various pumping intensities, are depicted in Fig. 7.10. Lasing was observed at 639 nm and a lasing threshold of ~400 pJ/pulse (~13 μJ/cm^2) was measured. This threshold pump power is only slightly higher compared to the value achieved in absence of PTAA, which is an indication of good growth of Alq_3 on the spin-coated PTAA layer. The fact that the presence of PTAA does not change the laser threshold pump power significantly compared to Alq_3 is also noticeable from Fig. 7.11, which compares the laser threshold for a set of samples comprising different thicknesses of the DCM_2-doped layer on PTAA to the threshold values of Fig. 7.9. Both sets of data coincide nicely.

In the above mentioned organic layer stacks, Alq_3 was used as the electron-transporting layer. This material, however, cannot be used in OLEDs with field-

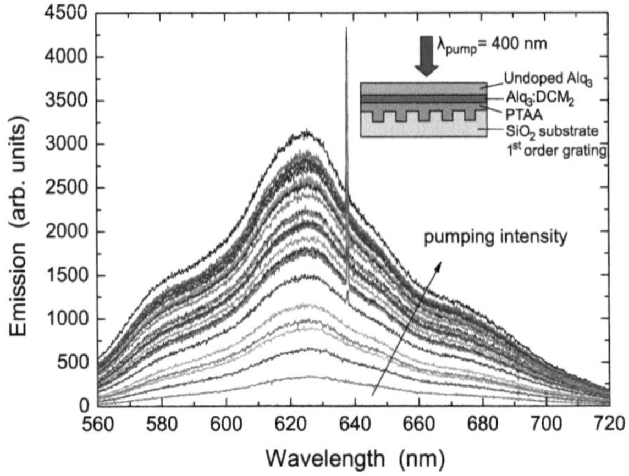

Fig. 7.10 Emission at room temperature from a lasing structure comprising a 50-nm thick PTAA hole-transporting layer on top of which 150 nm Alq_3:DCM_2 was deposited. Above the threshold a lasing peak appears in the spectrum. *Inset*: arrangement of the organic layer stack when comprising PTAA. (Measurements performed by IBM Research Zurich)

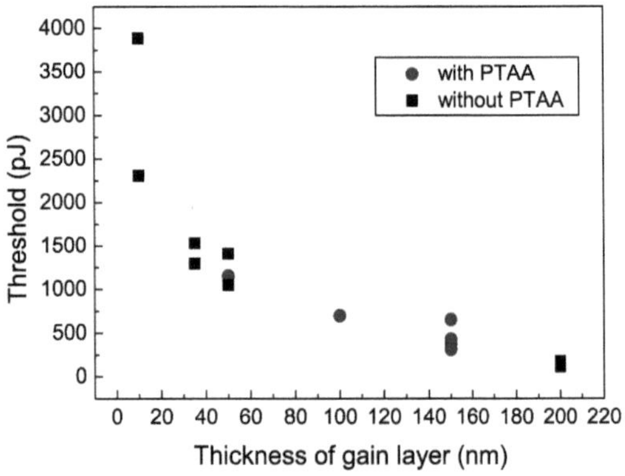

Fig. 7.11 Laser threshold values for a set of samples using different thicknesses of the DCM_2-doped gain layer on top of PTAA (*circles*) measured at room temperature. An increase of the laser threshold is observed for thinner gain layers. The threshold for devices without PTAA have been shown for comparison (*squares*). (Measurements performed by IBM Research Zurich)

effect electron transport, because it does not conduct electrons in a thin-film transistor configuration (Sect. 3.2). Indeed, in OLEDs with field-effect electron transport, electrons are transported towards the light-emission zone by field-effect. This field-effect electron transport occurs in an electron accumulation layer formed at the hetero-interface with the active light-emitting layer Alq_3:DCM_2. The non-planar molecular structure of Alq_3, however, prevents the formation of such an electron accumulation channel. Therefore, in order to optically pump organic layer stacks which are of direct use in OLEDs with field-effect electron transport, other electron-transporting materials have to be used. In this work, OLEDs with field-effect electron transport have been demonstrated using three different electron-transporting materials: PTCDI-$C_{13}H_{27}$, DFH-4T and DFHCO-4T. These three materials fulfil the two criteria that are important for device operation; they have a high electron field-effect mobility, and their LUMO is slightly lower than the LUMO of Alq_3 and DCM_2, making electron transport possible at the heterojunction between the electron-transporting layer and the light-emitting layer. In addition, DFH-4T and DFHCO-4T exhibit a larger optical gap than PTCDI-$C_{13}H_{27}$, which results in lower absorption in the red spectral region.

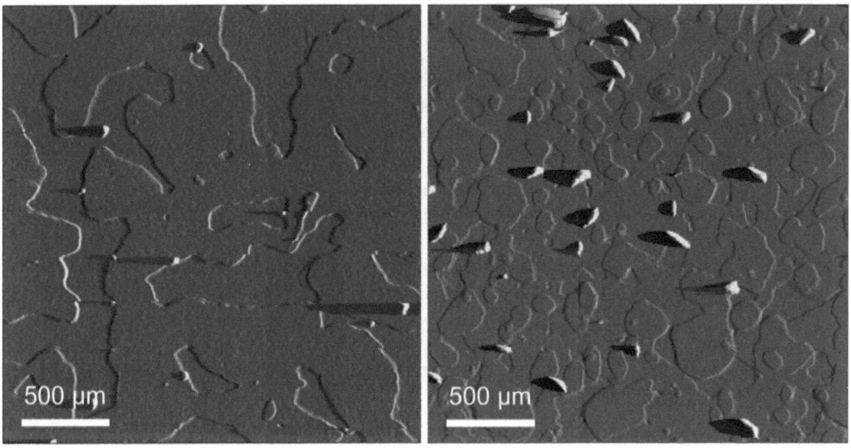

Fig. 7.12 2.5 μm × 2.5 μm AFM gradient image of (**a**) a 10-nm thick DFH-4T film deposited at 5 Å/s, while the substrate was at 60°C, and (**b**) a 10-nm thick DFHCO-4T film deposited at 5 Å/s and a substrate temperature of 70°C

Because of this low absorption in the red spectral region, the losses under optical pumping are expected to be lower for samples comprising DFH-4T and DFHCO-4T compared to samples comprising PTCDI-$C_{13}H_{27}$. However, also other factors, such as surface roughness and film morphology of the electron-transporting material may influence the laser threshold. As discussed in Chap. 4 (Sect. 4.1), thin DFH-4T and DFHCO-4T films deposited by vacuum thermal evaporation show Stransky-Krastanov growth, implying that the growth of the first few monolayers is two-dimensional, but beyond a certain thickness threshold the film starts to roughen strongly (3D growth mode). This rough growth of DFH-4T and DFHCO-4T might be problematic for achieving lasing and has to be reduced as much as possible. Deposition at high substrate temperatures and high deposition flux largely suppressed the 3D growth (Sect. 4.1). A further reduction of the amount of 3D features was possible by reducing the organic layer thickness. Note that a minimum thickness of about 10 nm was necessary in order to form an electron accumulation layer for electron transport. Figure 7.12 shows the AFM gradient images of a 10-nm thick DFH-4T and a 10-nm thick DFHCO-4T film grown on top of PαMS under optimized growth conditions. The 3D needle growth is strongly suppressed. Only for the DFHCO-4T film a few 3D features could still be observed.

Different organic layer stacks were then deposited on top of first order photonic feedback structures. These stacks used 50 nm PTAA, followed by the deposition of 50 nm Alq_3:DCM_2 and the deposition of 10 nm of the electron-transporting material (PTCDI-$C_{13}H_{27}$, DFH-4T or DFHCO-4T). To obtain a total thickness of 200 nm, 90 nm undoped Alq_3 was evaporated on top of these structures. This arrangement is schematically illustrated in Fig. 7.13. In addition, a sample comprising 180 nm Alq_3:DCM_2 on top of a 20-nm thick ITO layer was fabricated. This ITO layer was sputtered on a first order grating structure before deposition of the organic layer. For

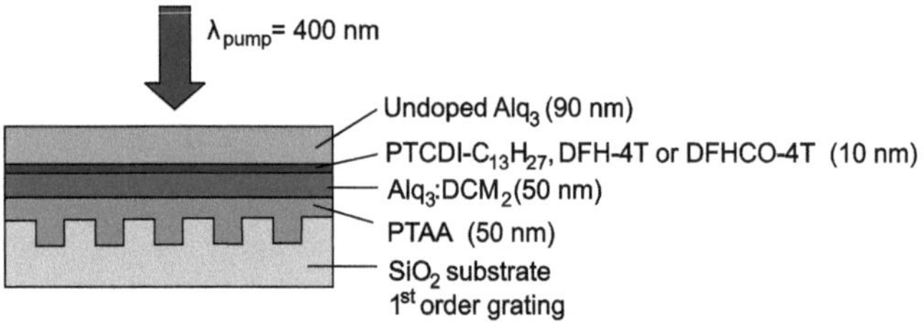

Fig. 7.13 Schematic illustration of the arrangement of the organic layer stack when comprising the electron-transporting layer PTCDI-$C_{13}H_{27}$, DFH-4T or DFHCO-4T

comparison, also an organic layer stack comprising 180 nm Alq_3:DCM_2 and 20 nm Alq_3 was fabricated.

Table 7.3 summarizes the measured average lasing thresholds of the different samples that were fabricated. For the structure comprising PTCDI-$C_{13}H_{27}$ as the electron-transporting material no lasing could be observed. This can be attributed to the substantial re-absorption of emitted photons by PTCDI-$C_{13}H_{27}$. The samples using DFH-4T and DFHCO-4T, on the other hand, did show lasing and an average lasing threshold of about 6000 pJ/pulse was measured for both structures. This observation is particularly encouraging because it indicates that optically pumped lasing can be achieved on material stacks including hole- and electron transporting layers which are of direct use in OLEDs with field-effect electron transport. The higher threshold of the DFH-4T and the DFHCO-4T samples compared to reference samples based on Alq_3 and with the same gain layer thickness is probably related to the layer morphology and the increased absorption at the pump wavelength ($\lambda = 400$ nm), since both materials should be transparent at the lasing wavelength ($\lambda \sim 640$ nm). Another interesting result is the fact that the presence of 20 nm ITO only slightly increased the lasing threshold. The lasing characteristics also were not modified due to the presence of ITO. This is in agreement with previous reports [182] and can be explained by the fact that 20 nm ITO introduces only very little absorption loss and does not significantly modify the confinement of the waveguide mode (Sect. 7.1.2).

From these measurements the exciton density at the laser threshold can be estimated. Taking into account a pump wavelength of 400 nm and a pump spot ra-

Table 7.3 Measured average lasing thresholds of different organic layer stacks deposited on first order gratings. Measurements were performed at room temperature using a pump wavelength of 400 nm

Sample	Average laser threshold
PTAA/Alq_3:DCM_2/PTCDI-$C_{13}H_{27}$/Alq_3	–
PTAA/Alq_3:DCM_2/DFH-4T/Alq_3	6188 pJ/pulse
PTAA/Alq_3:DCM_2/DFHCO-4T/Alq_3	6122 pJ/pulse
ITO/Alq_3:DCM_2	790 pJ/pulse
Alq_3/Alq_3:DCM_2	490 pJ/pulse

dius on the sample of approximately 25 μm, a lasing threshold of ~6100 pJ/pulse corresponds to a photon density of $6 \cdot 10^{14}$ cm^{-2} (or 300 μJ/cm^2) in the pump laser. Since the samples are optically pumped from the top, 35% of this incident energy is absorbed by the 90 nm thick undoped Alq_3 layer (Alq_3 absorption is $\alpha = 4.7 \cdot 10^4$ cm^{-1} at $\lambda = 400$ nm) and 10% by the 10 nm thick electron-transporting layer ($\alpha_{DFHCO-4T} = 7.5 \cdot 10^4$ cm^{-1} at $\lambda = 400$ nm). In addition, the Alq_3:DCM_2 gain layer absorbs only 20% of the energy at 400 nm excitation ($\alpha = 4.2 \cdot 10^4$ cm^{-1}). This leads to an exciton density at the laser threshold of $\sim 7 \cdot 10^{13}$ cm^{-2} for a device comprising an organic layer stack that is of direct use in an OLED with field-effect electron transport.

7.2.3 Discussion

As discussed in Sect. 7.1.1 the exciton density currently achieved under DC bias is estimated to be about $8 \cdot 10^9$ cm^{-2}. To bridge the gap of several orders of magnitude, clearly, improvements will have to come from both sides, an increase in achievable σ_{RZ} and a decrease of required σ_{RZ}. The former was discussed in Sect. 7.1.1. For the latter, the strong sensitivity of the lasing threshold power to details of the structure (Table 7.3) indicates that order-of-magnitude improvements could be achieved by reducing residual absorption, by further optimizing waveguide losses, and by studying feedback structures with higher quality factor. This latter proposition will be further discussed in the next paragraph.

7.3 Device with Integrated Field-Effect and Photonic Features

There are different possibilities to incorporate a photonic feedback structure in an OLED with field-effect electron transport. Four possible resonator/device geometries are illustrated in Fig. 7.14. Figure 7.14(a) displays the situation where feedback is generated in the vertical direction, by placing distributed Bragg reflectors underneath and on top of the structure. In this case, feedback is perpendicular to the surface. Another option, which increases the gain length considerably, is to provide feedback in the lateral plane. Light is then waveguided in the organic film via internal total reflection. As suggested by Fig. 7.14(b–d) incorporation of a lateral feedback mechanism can be performed in different ways.

A very important figure of merit for lasing is the exciton density at the laser threshold N_{th}. The optical resonator has an influence on this parameter through two important factors: the optical confinement factor Γ and the quality factor Q of the laser cavity. N_{th} is inversely proportional to both Γ and Q:

$$N_{th} \sim \frac{1}{\Gamma Q} \tag{7.2}$$

Γ corresponds to the electric light field energy contained inside the region with optical gain normalized to the total energy, whereas Q is inversely proportional to

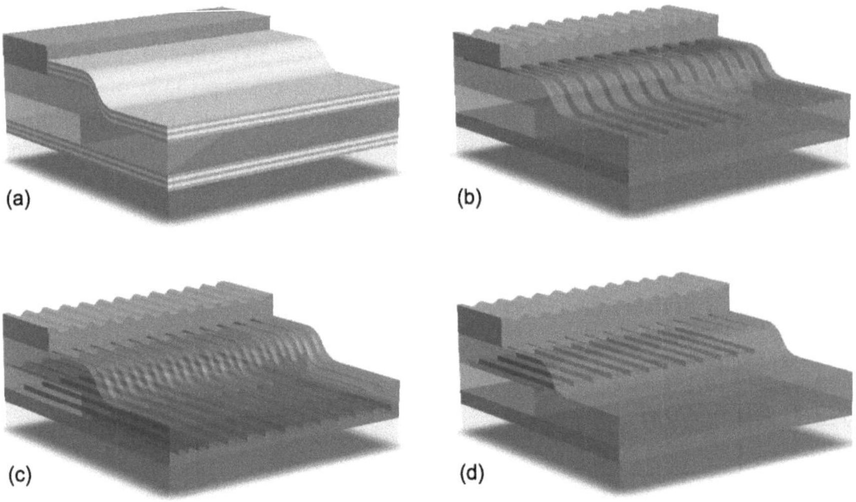

Fig. 7.14 Illustration of various possibilities to incorporate photonic feedback in an OLED with field-effect electron transport: (**a**) vertical resonator formed by DBR mirrors, (**b**) DFB structure imprinted in the organic layer stack, (**c**) DFB etched in the substrate before organic layer deposition, and (**d**) DFB structure etched in the insulator

the cavity loss coefficient and includes material absorption and output coupling from the resonator.

In close collaboration with IBM Research in Zurich, Γ and Q were calculated for, respectively, a vertically and a laterally oriented laser resonator structure, positioned around an organic layer stack comprising 120 nm PTAA, 20 nm Alq_3:DCM_2 and a 50-nm thick electron-transporting layer with low absorption losses such as DFH-4T or DFHCO-4T. The vertical resonator was formed by two DBR mirrors placed below and above the OLED with field-effect electron transport. Each of the 2×10 dielectric inorganic layer pairs were assumed to consist of 106 nm low refractive index material (SiO_2, $n = 1.47$) and 74 nm high refractive index material (Ta_2O_5[2], $n = 2.15$). For the laterally oriented laser resonator, one-dimensional feedback was provided by extended SiO_2 ridges, having a width of 106 nm and a height of 100 nm.

Table 7.4 compares the Γ (in the vertical direction) and Q values calculated for both photonic feedback geometries. Due to the limited gain layer thickness the resulting confinement factors are quite small. This seriously increases the laser threshold of the device (Eq. (7.2)), however, as demonstrated in Sect. 7.2.2 such thin gain layers do not exclude lasing. Note that Quantum-Well III-V lasers have similar confinement factors [334, 335]. The quality factor of both resonator geometries is also given in Table 7.4. For the vertical geometry the Q factor is the result of the normal cavity losses formed by 10 DBR layer pairs for output coupling. The nature of the first order feedback in the lateral direction, on the other hand, allows a nearly perfectly guided mode without output coupling and correspondingly a high Q factor

[2]Tantalum pentoxide.

Table 7.4 Comparison of the Γ and Q values of a vertical and lateral resonator geometry

	vertical resonator	lateral resonator
Confinement factor Γ (%)	5.3	3.8
Quality factor Q	2862	17071
Q with ITO	1190	702
Q with ITO and strong-absorbing ETL	78	203

is calculated. To assess the impact of the presence of ITO on the Q factor, similar calculations were performed assuming the organic layer stack is positioned on top of a 100 nm thick ITO layer. For the lateral feedback configuration this results in a dramatic reduction of the Q factor, while the vertical resonator geometry does not suffer as much. In contrast, the Q factor of the vertical resonator decreases more when a strong-absorbing electron-transporting layer such as PTCDI-$C_{13}H_{27}$ is used. Clearly, these results suggest that the incorporation of a lateral feedback mechanism in OLEDs with field-effect electron transport comprising low-absorbing materials such as DFH-4T or DFHCO-4T as the organic electron-transporting material and using only a very thin ITO layer as the bottom electrode might be a promising route to follow.

7.4 Summary and Conclusions

In this chapter, the prospects of realizing an electrically pumped laser using OLEDs with field-effect electron transport were investigated. By numerical simulations as well as by experiments it was demonstrated that OLEDs with field-effect electron transport allow a substantial reduction of the absorption losses thanks to the remoteness of the metallic cathode. In addition, very high current densities in the order of 13 A/cm^2 were measured without significant reduction of the external quantum efficiency. Based on these high current densities an exciton density per unit area of $8 \cdot 10^9$ cm^{-2} was calculated. Although this value is still several orders of magnitude lower than the values for the exciton density obtained from the threshold of optically pumped lasers, it is among the highest exciton densities per unit area reported to date for electrically pumped devices.

Amplified spontaneous emission measurements and optically pumped lasing experiments were performed on the host-guest system Alq_3:DCM_2 and on organic layer stacks including hole- and electron-transporting layers used in OLEDs with field-effect-assisted electron transport. Lasing could be achieved for gain layers of only 10 nm thick. However, the laser threshold was seriously increased for these thinner gain layers, which is in agreement with numerical simulations. It was found that the presence of the hole-transporting material PTAA does not change the laser threshold pump power significantly. Upon inserting electron-transporting materials, on the other hand, the laser threshold was increased about one order of magnitude. A thin layer of 20 nm ITO only slightly influenced the laser threshold. In the end, the different possibilities to incorporate a photonic feedback structure in OLEDs with field-effect electron transport were discussed. The incorporation of a lateral photonic feedback configuration in combination with the use of low absorbing electron-transporting materials and a thin ITO anode, might be a promising route towards the realization of an electrically pumped organic laser.

Chapter 8
General Conclusions and Future Outlook

8.1 Overview of the Main Results

This work focused on the design of new device and material concepts for organic light-emitting devices, thereby targeting high current densities and an improved control of the triplet concentration.

In Chap. 3, an OLED with field-effect electron transport was proposed as a new device architecture complementary to the list of existing electroluminescent devices. Compared to conventional OLEDs, the metallic top contact of this device is displaced by one to several micrometers from the light-emission zone. The injected electrons accumulate at an organic heterojunction and are transported towards the light-emission zone by field-effect. In this way, high mobilities could be achieved, enabling a high current density and a reduced number of charge carriers in the device. Light-emission was observed at a fixed position irrespective of the applied bias and smaller distances between the cathode and the insulator edge allowed for larger currents. The external quantum efficiency was found to be as high as for conventional OLEDs using the same materials. However, in contrast to standard OLEDs, the quantum efficiency of the new device was remarkably independent of the current, up to current densities of more than 10 A/cm^2. An increase of the external quantum efficiency could be obtained by decreasing the PTCDI-$C_{13}H_{27}$ layer thickness, indicating that re-absorption of emitted photons by PTCDI-$C_{13}H_{27}$ was limiting the device performance.

A further improvement of the device performance was the subject of Chap. 4. In that chapter, two diper fluorohexyl-quaterthiophene derivatives, DFH-4T and DFHCO-4T, were investigated as the electron-transporting material in OLEDs with field-effect electron transport. Both materials showed Stransky-Krastanov growth. The 3D needle growth could be largely suppressed by evaporation at high deposition rate and elevated substrate temperature. It was found out that DFHCO-4T top-contact transistors only function properly when using contacts with a low chemical reactivity such as Au and Ag, because these metals do not react with the organic semiconductor DFHCO-4T. Using a Au top-contact transistor configuration, mobilities as high as 4.6 cm^2/Vs were demonstrated. OLEDs with field-effect electron

DOI 10.1007/978-94-007-1608-7_8,

transport using DFH-4T or DFHCO-4T showed a higher external quantum efficiency compared to PTCDI-$C_{13}H_{27}$-based devices. This improved performance was attributed to the reduced light re-absorption in the red spectral region. A maximum external quantum efficiency of 0.25% was demonstrated for devices using DFH-4T as the electron-transporting material. Current densities up to 750 A/cm^2 in quasi-static operation were achieved in the electron channel of these devices.

In Chap. 5, two different approaches to control the triplet concentration in organic light-emitting devices were discussed. First, the pulsed excitation behavior of OLEDs with field-effect-assisted electron transport was investigated. Injection of electrons from the electron-transporting layer into the active light-emitting layer was found to determine the dependence of the light intensity on the applied pulse width. Depending on the applied voltage pulses up to 1 μs could be applied without affecting the light intensity. Since 1 μs is much shorter than the triplet exciton lifetime, pulsed excitation would reduce triplet accumulation in our device.

An alternative way to reduce the triplet concentration, namely the incorporation of a triplet scavenger, was discussed in the second part of Chap. 5. Using the non-classical triplet scavenger COT, employed before in liquid-phase organic dye lasers, nonvertical triplet energy transfer in solid conjugated polymer films was demonstrated for the first time. Thanks to its extraordinary large S_1–T_1 energy level splitting, COT was found to efficiently quench host phosphorescence without affecting the host fluorescence. In addition, owing to the relatively short lifetime of its low-lying relaxed triplet state, a fast deactivation of triplets was possible and consequently accumulation of triplet excitations could be prevented.

Polymer films incorporating the red-emitting metal-organic complex $Btp_2Ir(acac)$ were studied in Chap. 6. Spectral diffusion was observed at 10 wt% $Btp_2Ir(acac)$ dispersed in a polymer matrix, indicating that triplet exciton migration occurs at increased guest molecule concentration. This observation could explain the quenching by impurities like molecular oxygen in PS:(10 wt%)$Btp_2Ir(acac)$ films.

It was also found out that the decrease in Ph efficiency in CNPPP:$Btp_2Ir(acac)$ systems at increasing excitation intensity is dominated by mutual host-guest triplet-triplet annihilation. In the same films, a "delayed phosphorescence" was observed, which was attributed to a delayed production of guest triplets by recombination of trapped charge carriers on $Btp_2Ir(acac)$ guest molecules. The presence of hole traps formed by the $Btp_2Ir(acac)$ in the conjugated polymer CNPPP was evidenced by thermally stimulated luminescence measurements.

Additionally, the use of triplet-emitters as light-emitting species in the active layer of an organic laser were discussed. Three different metal-organic compounds were carefully investigated; however, no amplified spontaneous emission could be observed. Time-resolved pump-probe experiments identified an effective excited state absorption, which competes with radiative emission, as the main limitation.

In the last chapter, Chap. 7, the value of OLEDs with field-effect electron transport was discussed in the context of electrically pumped organic lasers. By numerical simulations as well as by experiments it was demonstrated that the new device concept allows a substantial reduction of the absorption losses thanks to the remoteness of the metallic cathode. In addition, high current densities were measured giving rise to high exciton densities per unit area. Optically pumped lasing experiments showed that lasing could be achieved for gain layers with a thickness of only 10 nm. Moreover, lasing was observed in organic layer stacks using hole-transporting and electron-transporting layers which are of direct use in OLEDs with field-effect electron transport. It was also demonstrated that a thin layer of 20 nm ITO only slightly influenced the laser threshold. The different possibilities to incorporate a photonic feedback structure in OLEDs with field-effect electron transport were discussed and it was found that combining low absorbing electron-transporting materials, a thin ITO anode and a lateral photonic feedback mechanism, might be a promising route towards the realization of an electrically pumped organic laser.

8.2 Suggestions for Further Research

In this work, new device and material concepts for organic light-emitting devices were suggested. The challenge is to optimize these concepts further and to use them for real applications. A first main limitation with respect to this are the organic materials available to date. Future work may therefore focus on the synthesis of new organic compounds, specific for high-brightness OLED and laser applications.

Nonvertical triplet acceptors, for example, exhibit properties which are very interesting for solid-state devices that suffer from triplet-state losses. However, so far this class of molecules was only used for liquid organic dye lasers. Because of this, most of the nonvertical compounds available today are liquid at room temperature, preventing the fabrication of solid-state devices. The synthesis of new nonvertical triplet acceptors which are solid at room temperature would therefore be very useful. In this way the effect of nonvertical triplet scavengers in electrically driven devices such as high-brightness OLEDs and other light-emitting structures could be investigated.

The synthesis of new organic materials would also be useful for further optimization of the performance of OLEDs with field-effect electron transport. New efficient red-emitting dye molecules or efficient near-infrared dyes that can be thermally evaporated might improve the external quantum efficiency of the device significantly. In addition, new high-mobility electron-transporting materials could result in higher current densities.

A further increase of the current density of the device could be obtained by using pulsed operation to higher voltages, by employing high-k dielectrics with low leakage current or by optimizing the device geometry. A confinement of the hole current could be obtained by etching narrow trenches or vias in the insulator layer. This would allow a significant increase of the hole current density, which is at this moment limiting the device performance. Along with this, the development of a technology to deposit the metallic top contact accurately centered around these vias might be very interesting. A circular integrated shadow mask could be used for this purpose.

Beyond that, the achievement of an electrically pumped organic laser remains a major challenge. In this work it has been shown that the reduced absorption losses and the high current densities obtained in OLEDs with field-effect electron transport make this device a promising configuration for laser applications. Different possibilities to incorporate a resonator structure in the device were suggested. Hence, the realization and the characterization of such an OLED with integrated field-effect and photonic feedback features would be very interesting. Future research may therefore focus on the development of a technology to fabricate these devices, thereby concentrating on techniques allowing a reduction of the ITO thickness, while maintaining a low resistivity. The fabricated devices could be measured and the effect of the feedback structure on the emission spectrum and the electrical characteristics could be investigated. Additionally, one may try to achieve lasing by applying higher electrical bias as well as by combining electrical and optical pumping. The latter technique might be very useful to determine how distant the purely electrically pumped organic laser still is.

Finally, also optical pumping of phosphorescent dye lasers might be subject of further research. Today, phosphorescent dye lasers have never been demonstrated, but one should keep in mind that efficient triplet-emitters have been synthesized relatively recently and that they were not intended for lasing applications. Synthesis of new efficient triplet-emitters with short radiative lifetime and low triplet-triplet absorption at the emission wavelength may therefore be interesting. These compounds could be investigated in solution at very low concentration and by using de-aerated solvents.

References

1. H. Mette, H. Pick, Z. Phys. **134**, 566 (1953)
2. M. Pope, H.P. Kallmann, P. Magnante, Electroluminescence in organic crystals. J. Chem. Phys. **38**, 2042 (1963)
3. R.G. Kepler, Charge carrier production and mobility in anthracene crystals. Phys. Rev. **119**(4), 1226–1229 (1960)
4. C.K. Chiang, C.R. Fincher, J.Y.W. Park, A.J. Heeger, H. Shirakawa, E.J. Louis, S.C. Gau, A.G. MacDiarmid, Electrical conductivity in doped polyacetylene. Phys. Rev. Lett. **39**(17), 1098–1101 (1977)
5. J.C. Kotz, P. Treichel, *Chemistry and Chemical Reactivity* (Saunders College Publishing, Philadelphia, 1999)
6. C. Kittel, *Introduction to Solid State Physics*, 6th edn. (Wiley, New York, 1986), p. 185
7. S. Verlaak, Small-molecule organic thin-film transistors: growth, charge transport and some applications. PhD thesis, K. U. Leuven (2004)
8. V.I. Arkhipov, V.A. Kolesnikov, A.I. Rudenko, Dispersive transport of charge-carriers in polycrystalline pentacene layers. J. Phys. D **17**(6), 1241–1254 (1984)
9. M. Pope, C.S. Swenberg, *Electronic Processes in Organic Crystals and Polymers*, 2nd edn. (Oxford University Press, New York, 1999)
10. V. Arkhipov, I. Fishchuk, A. Kadashchuck, H. Bässler, *Photophysics of Molecular Materials: From Single Molecules to Single Crystals* (Wiley, New York, 2006). Chap. 6: Charge transport in disordered organic semiconductors
11. W. Warta, N. Karl, Hot holes in naphthalene: high, electric-field-dependent mobilities. Phys. Rev. B **32**(2), 1172–1182 (1985)
12. N. Karl, J. Marktanner, Electron and hole mobilities in high purity anthracene single crystals. Mol. Cryst. Liq. Cryst. **355**, 149–173 (2001)
13. V. Podzorov, E. Menard, J.A. Rogers, M.E. Gershenson, Hall effect in the accumulation layers on the surface of organic semiconductors. Phys. Rev. Lett. **95**, 26601 (2005)
14. O. Ostroverkhova, D.G. Cooke, F.A. Hegmann, J.E. Anthony, V. Podzorov, M.E. Gershenson, O.D. Jurchescu, T.T.M. Palstra, Ultrafast carrier dynamics in pentacene, functionalized pentacene, tetracene, and rubrene single crystals. Appl. Phys. Lett. **88**, 162101 (2006)
15. P.G.L. Comber, W.E. Spear, Electronic transport in amorphous silicon films. Phys. Rev. Lett. **25**(8), 509–511 (1970)
16. G. Horowitz, R. Hajlaoui, P. Delannoy, Temperature-dependence of the field-effect mobility of sexithiophene—determination of the density of traps. J. Phys. **5**(4), 355–371 (1995)
17. M. Mottaghi, G. Horowitz, Field-induced mobility degradation in pentacene thin-film transistors. Org. Electron. **7**, 528–536 (2006)
18. H. Bässler, Charge transport in disordered organic photoconductors: a Monte-Carlo simulation study. Phys. Status Solidi B **175**(1), 15–56 (1993)

S. Schols, *Device Architecture and Materials for Organic Light-Emitting Devices*,
DOI 10.1007/978-94-007-1608-7, © Springer Science+Business Media B.V. 2011

19. M.C.J.M. Vissenberg, M. Matters, Theory of the field-effect mobility in amorphous organic transistors. Phys. Rev. B **57**(20), 12964–12967 (1998)
20. J.A. Barltrop, J.D. Coyle, *Excited States in Organic Chemistry* (Wiley-VCH, New York, 1975)
21. M.V. der Auweraer, Photophysics and photochemistry of molecular materials, B-KUL-G0I12A
22. Y. Shi, J. Liu, K. Han, Investigation of the internal conversion time of the chlorophyll a from S_3, S_2 to S_1. Chem. Phys. Lett. **410**(4–6), 260–263 (2005)
23. C.W. Tang, S.A. VanSlyke, Organic electroluminescent diodes. Appl. Phys. Lett. **51**(12), 913–915 (1987)
24. L.S. Hung, C.H. Chen, Recent progress of molecular organic electroluminescent materials and devices. Mater. Sci. Eng. R. **39**, 143–222 (2002)
25. A. Bernanose, M. Comte, P. Vouaux, A new method of light emission by certain organic compounds. J. Chim. Phys. **50**, 64 (1953)
26. A. Bernanose, P. Vouaux, Organic electroluminescence type of emission. J. Chim. Phys. **50**, 261 (1953)
27. J.H. Burroughes, D.D. Bradley, A.R. Brown, R.N. Marks, K. Mackay, R.H. Friend, P.L. Burnst, A.B. Holmes, Light-emitting diodes based on conjugated polymers. Nature **347**, 539 (1990)
28. R.H. Friend, R.W. Gymer, A.B. Holmes, J.H. Burroughes, R.N. Marks, C. Taliani, D.D.C. Bradley, D.A.D. Santos, J.L. Brédas, M. Lögdlund, W.R. Salaneck, Electroluminescence in conjugated polymers. Nature **397**, 121 (1999)
29. F. So, J. Kido, P. Burrows, Organic light-emitting devices for solid-state lighting. Mater. Res. Soc. Bull. **33**, 663–669 (2008)
30. S.A. VanSlyke, C.W. Tang, Organic electroluminescent devices having improved power conversion efficiencies. US patent 4,539,507
31. P.W.M. Blom, M.J.M. de Jong, Electrical characterization of polymer light-emitting diodes. IEEE J. Sel. Top. Quantum Electron. **4**(1), 105–112 (1998)
32. C. Hosokawa, H. Tokailin, H. Higashi, T. Kusumoto, Transient behavior of organic thin film electroluminescence. Appl. Phys. Lett. **60**(10), 1220–1222 (1992)
33. J. Kido, Y. Lizumi, Fabrication of highly efficient organic electroluminiscent devices. Appl. Phys. Lett. **73**(19), 2721–2723 (1998)
34. C. Adachi, M.A. Baldo, M.E. Thompson, S.R. Forrest, Nearly 100% internal phosphorescence efficiency in an organic light emitting device. J. Appl. Phys. **90**(10), 5048–5051 (2001)
35. N.K. Patel, S. Cina, J.H. Burroughes, High-efficiency organic light-emitting diodes. IEEE J. Quantum Electron. **8**(2), 346–361 (2002)
36. M.A. Baldo, D.F. O'Brien, M.E. Thompson, S.R. Forrest, Excitonic singlet-triplet ratio in a semiconducting organic thin film. Phys. Rev. B **60**(20), 14422–14428 (1999)
37. M.A. Baldo, D.F. O'Brien, Y. You, A. Shoustikov, S. Sibley, M.E. Thompson, S.R. Forrest, Highly efficient phosphorescent emission from organic electroluminescent devices. Nature **395**, 151–154 (1998)
38. D. Tanaka, H. Sasabe, Y. Li, S. Su, T. Takeda, J. Kido, Ultra high efficiency green organic light-emitting devices. Jpn. J. Appl. Phys. **46**(1), L10–L12 (2007)
39. D. Tanaka, Y. Agata, T. Takeda, S. Watanabe, J. Kido, High luminous efficiency blue organic light-emitting devices using high triplet excited energy materials. Jpn. J. Appl. Phys. **46**(5), L117–L119 (2007)
40. O. Inganäs, T. Granlund, M. Theander, M. Berggren, M.R. Andersson, A. Ruseckas, V. Sundström, Optical emission from confined poly(thiophene) chains. Opt. Mater. **9**, 104–108 (1998)
41. C.W. Tang, S.A. VanSlyke, C.H. Chen, Electroluminescence of doped organic thin films. J. Appl. Phys. **65**(9), 3610–3616 (1989)
42. Y. Lin, W. Chou, S. Lin, Enhanced efficiency in polymer light-emitting diodes due to the improvement of charge-injection balance. Appl. Phys. Lett. **88**, 071108 (2006)

43. T. Yamasaki, K. Sumioka, T. Tsutsui, Organic light-emitting device with an ordered monolayer of silica microspheres as a scattering medium. Appl. Phys. Lett. **76**(10), 1243–1245 (2000)
44. G. Parthasarathy, P.E. Burrows, V. Khalfin, V.G. Kozlov, S.R. Forrest, A metal-free cathode for organic semiconductor devices. Appl. Phys. Lett. **72**(17), 2138–2140 (1998)
45. F. Dinelli, M. Mugia, P. Levy, M. Cavallini, F. Biscarini, Spatially correlated charge transport in organic thin film transistors. Phys. Rev. Lett. **92**(11), 116802 (2004)
46. C. Tanase, E.J. Meijer, P.W.M. Blom, D.M. de Leeuw, Unification of the hole transport in polymeric field-effect transistors and light-emitting diodes. Phys. Rev. Lett. **91**, 216601 (2003)
47. A. Dodabalapur, H.E. Katz, L. Torsi, Molecular orbital energy level engineering in organic transistors. Adv. Mater. **8**(10), 853–855 (1996)
48. A. Hepp, H. Heil, W. Weise, M. Ahles, R. Schmechel, H. von Seggern, Light-emitting field-effect transistor based on a tetracene thin film. Phys. Rev. Lett. **91**(15), 157406 (2003)
49. M. Muccini, A bright future for organic field-effect transistors. Nat. Mater. **5**, 605–613 (2006)
50. S. Verlaak, D. Cheyns, M. Debucquoy, V. Arkhipov, P. Heremans, Numerical simulation of tetracene light-emitting transistors: a detailed balance of exciton processes. Appl. Phys. Lett. **85**(12), 2405–2407 (2004)
51. J. Reynaert, D. Cheyns, D. Janssen, R. Müller, V.I. Arkhipov, J. Genoe, G. Borghs, P. Heremans, Ambipolar injection in a submicron-channel light-emitting tetracene transistor with distinct source and drain contacts. J. Appl. Phys. **97**, 114501 (2005)
52. C. Santato, R. Capelli, M.A. Loi, M. Murgia, F. Cicoira, V.A.L. Roy, P. Stallinga, R. Zamboni, C. Rost, S.F. Karg, M. Muccini, Tetracene-based organic light-emitting transistors: optoelectronic properties and electron injection mechanism. Synth. Met. **146**, 329–334 (2004)
53. T. Sakanoue, E. Fujiwara, R. Yamada, H. Tada, Visible light emission from polymer-based field-effect transistors. Appl. Phys. Lett. **84**(16), 3037–3039 (2004)
54. M. Ahles, A. Hepp, R. Schmechel, H. von Seggern, Light emission from a polymer transistor. Appl. Phys. Lett. **84**(3), 428–430 (2004)
55. J. Swensen, D. Moses, A.J. Heeger, Light emission in the channel region of a polymer thin-film transistor fabricated with gold and aluminum for the source and drain electrodes. Synth. Met. **153**, 53–56 (2005)
56. H. Nakanotani, S. Akiyama, D. Ohnishi, M. Moriwake, M. Yahiro, T. Yoshihara, S. Tobita, C. Adachi, Extremely low-threshold amplified spontaneous emission of 9,9'-spirobifluorene derivates and electroluminescence from field-effect transistor structure. Adv. Funct. Mater. **17**, 2328–2335 (2007)
57. T. Oyamada, H. Uchiuzou, S. Akiyama, Y. Oku, N. Shimoji, K. Matsushige, H. Sasabe, C. Adachi, Lateral organic light-emitting diode with field-effect transistor characteristics. J. Appl. Phys. **98**, 074506 (2005)
58. C. Rost, S. Karg, W. Riess, M.A. Loi, M. Murgia, M. Muccini, Ambipolar light-emitting organic field-effect transistor. Appl. Phys. Lett. **85**(9), 1613–1615 (2004)
59. M.A. Loi, C. Rost, M. Murgia, S. Karg, W. Riess, M. Muccini, Tuning optoelectronic properties of ambipolar organic light-emitting transistor using a bulk-heterojunction approach. Adv. Funct. Mater. **16**, 41–47 (2006)
60. C. Rost, S. Karg, W. Riess, M.A. Loi, M. Murgia, M. Muccini, Light-emitting ambipolar organic heterostructure field-effect transistor. Synth. Met. **146**, 237–241 (2004)
61. R. Capelli, F. Dinelli, M.A. Loi, M. Murgia, R. Zamboni, M. Muccini, Ambipolar organic light-emitting transistors employing heterojunctions of n-type and p-type materials as the active layer. J. Phys., Condens. Matter **18**, S2127–S2138 (2006)
62. F. Dinelli, R. Capelli, M.A. Loi, M. Murgia, M. Muccini, A. Facchetti, T.J. Marks, High-mobility ambipolar transport in organic light-emitting transistors. Adv. Mater. **18**, 1416–1420 (2006)
63. E.B. Namdas, P. Ledochowitsch, J.D. Yuen, D. Moses, A.J. Heeger, High performance light emitting transistors. Appl. Phys. Lett. **92**, 183304 (2008)

64. S.D. Vusser, S. Steudel, S. Schols, S. Verlaak, J. Genoe, W.D. Oosterbaan, L. Lutsen, D. Vandezande, P. Heremans, A light-emitting organic field-effect transistor using an organic heterostructure inside the transistor channel. Appl. Phys. Lett. **89**, 223504 (2006)
65. J. Zaumseil, R. Friend, H. Sirringhaus, Spatial control of the recombination zone in an ambipolar light-emitting organic transistor. Nat. Mater. **5**(1), 69–74 (2006)
66. J.S. Swensen, C. Soci, A.J. Heeger, Light emission from an ambipolar semiconducting polymer field-effect transistor. Appl. Phys. Lett. **87**, 253511 (2005)
67. J. Zaumseil, C.L. Donley, J. Kim, R.H. Friend, H. Sirringhaus, Efficient top-gate, ambipolar, light-emitting field-effect transistors based on a green-light-emitting polyfluorene. Adv. Mater. **18**, 2708–2712 (2006)
68. J.S. Swensen, J. Yuen, D. Gargas, S.K. Buratto, A.J. Heeger, Light emission from an ambipolar semiconducting polymer field effect transistor: Analysis of the device physics. J. Appl. Phys. **102**, 013103 (2007)
69. R.C.G. Naber, M. Bird, H. Sirringhaus, A gate dielectric that enables high ambipolar mobilities in polymer light-emitting field-effect transistors. Appl. Phys. Lett. **93**, 023301 (2008)
70. T. Sakanoue, M. Yahiro, C. Adachi, H. Uchiuzou, T. Takahashi, A. Toshimitsu, Ambipolar light-emitting organic field-effect transistors using a wide-band-gab blue-emitting small molecule. Appl. Phys. Lett. **90**, 171118 (2007)
71. E.C.P. Smits, S. Setayesh, T.D. Anthopoulos, M. Buechel, W. Nijssen, R. Coehoorn, P.W.M. Blom, B. de Boer, D.M. de Leeuw, Near-infrared light-emitting ambipolar organic field-effect transistors. Adv. Mater. **19**, 734–738 (2007)
72. D.L. Smith, P.P. Ruden, Analytic device model for light-emitting ambipolar organic semiconductor field-effect transistors. Appl. Phys. Lett. **89**, 233519 (2006)
73. M. Kemerink, D.S.H. Charrier, E.C.P. Smits, S.G.J. Mathijssen, D.M. de Leeuw, R.A.J. Janssen, On the width of the recombination zone in ambipolar organic field effect transistors. Appl. Phys. Lett. **93**, 033312 (2008)
74. T. Takahashi, T. Takenobu, J. Takeya, Y. Iwasa, Ambipolar light-emitting transistors of a tetracene single crystal. Adv. Funct. Mater. **17**, 1623–1628 (2007)
75. T. Takenobu, S.Z. Bisri, T. Takahashi, M. Yahiro, C. Adachi, Y. Iwasa, High current density in light-emitting transistors of organic single crystals. Phys. Rev. Lett. **100**, 066601 (2008)
76. H. Nakanotani, R. Kabe, M. Yahiro, T. Takenobu, Y. Iwasa, C. Adachi, Blue-light-emitting ambipolar field-effect transistors using an organic single crystal of 1,4-bis(4-methylstyryl)benzene. Appl. Phys. Express **1**, 091801 (2008)
77. J. Zaumseil, H. Sirringhaus, Electron and ambipolar transport in organic field-effect transistors. Chem. Rev. **107**, 1296–1323 (2007)
78. F. Cicoira, C. Santato, Organic light emitting field effect transistors: advances and perspectives. Adv. Funct. Mater. **17**, 3421–3434 (2007)
79. T.P.I. Saragi, R. Pudzich, T. Fuhrmann, J. Salbecka, Organic phototransistor based on intramolecular charge transfer in a bifunctional spiro compound. Appl. Phys. Lett. **84**(13), 2334–2336 (2004)
80. M. Debucquoy, S. Verlaak, S. Steudel, K. Myny, J. Genoe, P. Heremans, Correlation between bias stress instability and phototransistor operation of pentacene thin-film transistors. Appl. Phys. Lett. **91**, 103508 (2007)
81. T.N. Ng, W.S. Wong, M.L. Chabinyc, S. Sambandan, R.A. Street, Flexible image sensor array with bulk heterojunction organic photodiode. Appl. Phys. Lett. **92**, 213303 (2008)
82. C.J. Brabec, N.S. Sariciftci, J.C. Hummelen, Plastic solar cells. Adv. Funct. Mater. **11**(1), 15–26 (2001)
83. H. Sirringhaus, N. Tessler, R.H. Friend, Integrated optoelectronic devices based on conjugated polymers. Science **280**, 1741–1744 (1998)
84. N. Tessler, Laser based on semiconducting organic materials. Adv. Mater. **11**(5), 363–370 (1999)
85. M.D. McGehee, A.J. Heeger, Semiconducting (conjugated) polymers as material for solid-state lasers. Adv. Mater. **12**(22), 1655–1668 (2000)
86. I.D.W. Samuel, G.A. Turnbull, Organic semiconductor lasers. Chem. Rev. **107**, 1272–1295 (2007)

87. T.H. Maiman, Stimulated optical radiation in ruby. Nature **187**(4736), 493–494 (1960)
88. D. Meschede, *Optics, Light and Lasers: The Practical Approach to Modern Aspects of Photonics and Laser Physics* (Wiley-VCH, Weinheim, 2004)
89. J. Engelen, Optische communicatie h245 & h806, k.u.leuven, ir.elektronica: ict-telecommunicatie en telematica (2003)
90. V.G. Kozlov, V. Bulovic, P.E. Burrows, S.R. Forrest, Laser action in organic semiconductor waveguide and double-heterostructure devices. Nature **389**, 362–364 (1997)
91. G. Kranzelbinder, G. Leising, Organic solid-state lasers. Rep. Prog. Phys. **63**(5), 729–762 (2000)
92. B. Schweitzer, G. Wegmann, H. Giessen, D. Hertel, H. Bässler, R.F. Mahrt, U. Scherf, K. Müllen, The optical gain mechanism in solid conjugated polymers. Appl. Phys. Lett. **72**, 2933–2935 (1998)
93. A.R. Brown, A. Pomp, C.M. Hart, D.M. de Leeuw, Logic gates made from polymer transistors and their use in ring oscillators. Science **270**(5238), 972–974 (1995)
94. Z. Bao, A. Bodabalapur, A.J. Lovinger, Soluble and processable regioregular poly(3-hexylthiophene) for thin film field-effect transistor applications with high mobility. Appl. Phys. Lett. **69**(26), 4108–4110 (1996)
95. C.D. Dimitrakopoulos, P.R.L. Malenfant, Organic thin film transistors for large area electronics. Adv. Mater. **14**(2), 99–117 (2002)
96. V.G. Kozlov, V. Bulovic, P.E. Burrows, M. Baldo, V.B. Khalfin, G. Parthasarathy, S.R. Forrest, Study of lasing action based on förster energy transfer in optically pumped organic semiconductor thin films. J. Appl. Phys. **84**(8), 4096–4108 (1998)
97. D. Schneider, T. Rabe, T. Riedl, T. Dobbertin, M. Kröger, E. Becker, H. Johannes, W. Kowalsky, T. Weimann, J. Wang, P. Hinze, Ultrawide tuning range in doped organic solid-state lasers. Appl. Phys. Lett. **85**(11), 1886–1888 (2004)
98. D. Schneider, T. Rabe, T. Riedl, T. Dobbertin, O. Werner, M. Kröger, E. Becker, H. Johannes, W. Kowalsky, T. Weimann, J. Wang, P. Hinze, A. Gerhard, P. Stössel, H. Vestweber, Deep blue widely tunable organic solid-state laser based on a spirobifluorene derivative. Appl. Phys. Lett. **84**(23), 4693–4695 (2004)
99. T. Riedl, T. Rabe, H.-H. Johannes, W. Kowalsky, J. Wang, T. Weimann, R. Hinze, B. Nehls, T. Farrell, U. Scherf, Tunable organic thin-film laser pumped by an inorganic violet diode laser. Appl. Phys. Lett. **88**, 241116 (2006)
100. G. Wegmann, H. Giessen, A. Greiner, R.F. Mahrt, Laser emission from a solid conjugated polymer: gain, tunability and coherence. Phys. Rev. B **57**(8), R4218–R4221 (1998)
101. A.K. Sheridan, G.A. Turnbul, A.N. Safonov, I.D.W. Samuel, Tuneability of amplified spontaneous emission through control of the waveguide-mode structure in conjugated polymer films. Phys. Rev. B **62**(18), R11929–R11932 (2000)
102. G.A. Turnbull, T.F. Krauss, W.L. Barnes, I.D.W. Samuel, Tuneable distributed feedback lasing in MEH-PPV films. Synth. Met. **121**, 1757–1758 (2001)
103. S. Riechel, U. Lemmer, J. Feldmann, S. Berleb, A.G. Mückl, W. Brütting, Very compact tunable solid-state laser utilizing a thin-film organic semiconductor. Opt. Lett. **26**(9), 593–595 (2001)
104. B. Schütte, H. Gothe, S.I. Hintschich, M. Sudzius, H. Fröb, V.G. Lyssenko, K. Leo, Continuously tunable laser emission from a wedge-shaped organic microcavity. Appl. Phys. Lett. **92**, 163309 (2008)
105. D. Amarasinghe, A. Ruseckas, A.E. Vasdekis, M. Goossens, G.A. Turnbull, I.D.W. Samuel, Broadband solid state optical amplifier based on a semiconducting polymer. Appl. Phys. Lett. **89**, 201119 (2006)
106. J.R. Lawrence, G.A. Turnbull, I.D.W. Samuel, Broadband optical amplifier based on a conjugated polymer. Appl. Phys. Lett. **80**(17), 3036–3038 (2002)
107. I.D.W. Samuel, G.A. Turnbull, Polymer lasers: recent advances. Mater. Today 28–35 (2004)
108. U. Rauscher, H. Bässler, D.D.C. Bradley, M. Hennecke, Exciton versus band description of the absorption and luminescence spectra in poly(p-phenylenevinylene). Phys. Rev. B **42**(16), 9830–9836 (1990)

109. R. Kersing, U. Lemmer, R.F. Mahrt, K. Leo, H. Kurz, H. Bässler, E.O. Böbel, Femtosecond energy relaxation in π-conjugated polymers. Phys. Rev. B **70**(24), 3820–3823 (1993)
110. M. Andersson, G. Yub, A. Heeger, Photoluminescence and electroluminescence of films from soluble PPV-polymers. Synth. Met. **85**, 1275–1276 (1997)
111. S. Tasch, A. Niko, G. Leising, U. Scherf, Highly efficient electroluminescence of new wide band gap ladder-type poly(para-phenylenes). Appl. Phys. Lett. **68**(8), 1090–1092 (1996)
112. B.R. Hsieh, Y. Yu, E.W. Forsythe, G.M. Schaaf, W.A. Feld, A new family of highle emissive soluble poly(p-phenylene vinylene) derivatives. A step toward fully conjugated blue-emitting poly(p-phenylene vinylenes). J. Am. Chem. Soc. **120**, 231–232 (1998)
113. L. Chan, Y. Lee, C. Chen, Synthesis and characterization of 3,4-diphenylmaleimide copolymers that exhibit organic to red photoluminescence and electroluminescence. Macromolecules **39**, 3262–3269 (2006)
114. M. Berggren, A. Dodabalapur, R.E. Slusher, Stimulated emission and lasing in dye-doped organic thin films with förster transfer. Appl. Phys. Lett. **71**, 2230 (1997)
115. A. Tsumara, H. Koezuka, T. Ando, Macromolecular electronic device: field-effect transistor with a polythiophene thin film. Appl. Phys. Lett. **49**(18), 1210–1212 (1986)
116. V.L. Broude, V.S. Mashkevich, A.F. Prikhot'ko, N.F. Prokopyuk, M.S. Soskin, On the possibility of stimulated emission in systems with electronic vibrational levels. Fiz. Tverd. Tela **4**, 2976 (1962)
117. P.P. Sorokin, J.R. Lankard, Stimulated emission observed from an organic dye chloro-aluminum phthalocyanine. IBM J. Res. Dev. **10**, 162–163 (1966)
118. M. Maeda, *Laser Dyes: Properties of Organic Compounds for Dye Lasers* (OHM, Tokyo, 1984)
119. B.H. Soffer, B.B. McFarland, Continously tunable, narrow-band organic dye lasers. Appl. Phys. Lett. **10**(10), 266–267 (1967)
120. N. Karl, Laser emission form an organic molecular crystal. Phys. Status Solidi A **13**, 651 (1972)
121. O.S. Avanesjan, V.A. Benderskii, V.K. Brikenstein, V.L. Broude, L.I. Korshunov, A.G. Lavrushko, I.I. Tartakovskii, Anthracene crystals under intensive optical pumping. Mol. Cryst. Liq. Cryst. **29**, 165–174 (1974)
122. H. Kogelnik, C.V. Shank, Stimulated emission in a periodic structure. Appl. Phys. Lett. **18**(4), 152–154 (1971)
123. T.W. Hänsch, M. Pernier, A.L. Schawlow, Laser action of dyes in gelatin. IEEE J. Quantum Electron. **7**(1), 45–46 (1971)
124. F.P. Schäfer, K.H. Drexhage, *Dye Lasers*, vol. 1, 2nd rev. edn. (Springer, Berlin, 1977)
125. D. Zhang, Y. Wang, D. Ma, Random lasing emission from a red fluorescent dye doped polystyrene film containing dispersed polystyrene nanoparticles. Appl. Phys. Lett. **91**, 091115 (2007)
126. D. Moses, High quantum efficiency luminescence from a conducting polymer in solution: A novel polymer laser dye. Appl. Phys. Lett. **60**(26), 3215–3216 (1992)
127. L.J. Rothberg, M. Yan, F. Papadimitrakopoulos, M.E. Galvin, E.W. Kwock, T.M. Miller, Photophysics of phenylenevinylene polymers. Synth. Met. **80**, 41–58 (1996)
128. L.J. Rothberg, M. Yan, S. Son, M.E. Galvin, E.W. Kwock, T.M. Miller, H.E. Katz, R.C. Haddon, F. Papadimitrakopoulos, Intrinsic and extrinsic constraints on phenylenevinylene polymer electroluminescence. Synth. Met. **78**, 213–236 (1996)
129. G.J. Denton, N. Tessler, N.T. Harrison, R.H. Friend, Factors influencing stimulated emission from poly(p-phenylenevinylene). Phys. Rev. Lett. **78**(4), 733–736 (1997)
130. F. Hide, B.J. Schwartz, M.A. Diaz-Garcia, A.J. Heeger, Laser emission from solutions and films containing semiconducting polymer and titanium dioxide nanocrystals. Chem. Phys. Lett. **256**, 424–430 (1996)
131. W. Graupner, G. Leising, G. Lanzani, M. Nisoli, S.D. Silvestri, U. Scherf, Femtosecond relaxation of photoexcitations in a poly(para-phenylene)-type ladder polymer. Phys. Rev. Lett. **76**(5), 847–850 (1996)
132. N. Tessler, G.J. Denton, R.H. Friend, Lasing from conjugated polymer microcavities. Nature **382**, 695–697 (1996)

133. F. Hide, M.A. Diaz-Garcia, B.J. Schwartz, M.R. Andersson, Q. Pei, A.J. Heeger, Semiconducting polymers: a new class of solid-state laser materials. Science **273**, 1833–1836 (1996)
134. S.V. Frolov, M. Ozaki, W. Gellermann, Z.V. Vareeny, K. Yoshino, Mirrorless lasing in conducting polymer poly(2,5-dioctyloxy-p-phenylenevinylene) films. J. Appl. Phys. **35**(10B), 1371–1373 (1996)
135. M. Berggren, A. Dodabalapur, R.E. Slusher, Z. Bao, Light amplification in organic thin films using cascade energy transfer. Nature **389**, 466–469 (1997)
136. V.G. Kozlov, G. Parthasarathy, P.E. Burrows, S.R. Forrest, Y. You, M.E. Thompson, Optically pumped blue organic semiconductor lasers. Appl. Phys. Lett. **72**(2), 144–146 (1998)
137. T. Aimono, Y. Kawamura, K. Goushi, H. Yamamoto, H. Sasebae, C. Adachi, 100% fluorescence efficiency of 4.4'-bis[(n-carbazole)styryl]biphenyl in a solid film and the very low amplified spontanous emission threshold. Appl. Phys. Lett. **86**(7), 071110 (2005)
138. M.D. McGehee, R. Gupta, S. Veenstra, E.K. Miller, M.A. Diaz-Garcia, A.J. Heeger, Amplified spontaneous emission from photopumped films of a conjugated polymer. Phys. Rev. B **58**(11), 7035–7039 (1998)
139. A. Andreev, F. Quochi, F. Cordella, A. Mura, G. Bongiovanni, H. Sitter, G. Hlawacek, C. Teichert, N.S. Sariciftci, Coherent random lasing in the deep blue from self-assembled organic nanofibers. J. Appl. Phys. **99**, 034305 (2006)
140. D. Schneider, T. Rabe, T. Riedl, T. Dobbertin, M. Kröger, E. Becker, H. Johannes, W. Kowalsky, T. Weimann, J. Wang, P. Hinze, Laser threshold reduction in an all-spiro guest-host system. Appl. Phys. Lett. **85**(10), 1659–1661 (2004)
141. T. Rabe, M. Hoping, D. Schneider, E. Becker, H. Johannes, W. Kowalsky, T. Weimann, J. Wang, P. Hinze, B.S. Nehls, U. Scherf, T. Farrell, T. Riedl, Threshold reduction in polymer lasers based on poly(9,9-dioctylfluorene) with statistical binaphthyl units. Adv. Funct. Mater. **12**, 1188–1192 (2005)
142. C. Zenz, W. Graupner, S. Tasch, G. Leising, K. Müllen, U. Scherf, Blue green stimulated emission from a high gain conjugated polymer. Appl. Phys. Lett. **71**(18), 2566–2568 (1997)
143. A. Dodabalapur, M. Berggren, R.E. Slusher, Z. Bao, A. Timko, P. Schiortino, E. Laskowsk, H.E. Katz, O. Nalamasu, Resonators and materials for organic lasers based on energy transfer. IEEE J. Quantum Electron. **4**(1), 67–74 (1998)
144. G. Heliotis, R. Xia, D.D.C. Bradley, G.A. Turnbull, I.D.W. Samuel, P. Andrew, W.L. Barnes, Two-dimensional distributed feedback lasers using a broadband, red polyfluorene gain medium. J. Appl. Phys. **96**(12), 6959–6965 (2004)
145. C. Bauer, H. Giessen, B. Schnabel, E. Kley, C. Schmitt, U. Scherf, R.F. Mahrt, A surface-emitting circular grating polymer laser. Adv. Mater. **13**(15), 1161–1164 (2001)
146. C. Karnutsch, C. Gärtner, V. Haug, U. Lemmer, T. Farrell, B.S. Nehls, U. Scherf, J. Wang, T. Weimann, G. Heliotis, C. Pflumm, J.C. deMello, D.D.C. Bradley, Low threshold blue conjugated polymer lasers with first-and second-order distributed feedback. Appl. Phys. Lett. **89**, 201108 (2006)
147. K. Baumann, T. Stöferle, N. Moll, R. Mahrt, T. Wahlbrink, J. Bolten, T. Mollenhauser, C. Moormann, U. Scherf, Organic mixed-order photonic crystal lasers with ultrasmall footprint. Appl. Phys. Lett. **91**, 171108 (2007)
148. C. Karnutsch, C. Pflumm, G. Heliotis, J.C. deMello, D.D.C. Bradley, Improved organic semiconductor lasers based on a mixed-order distributed feedback resonator design. Appl. Phys. Lett. **90**, 131104 (2007)
149. U. Scherf, S. Riechel, U. Lemmer, R.F. Mahrt, Conjugated polymers: lasing and stimulated emission. Curr. Opin. Solid State Mater. Sci. **5**, 143–154 (2001)
150. A. Haugeneder, M. Neges, C. Kallinger, W. Spirkl, U. Lemmer, J. Feldmann, M.-C. Amann, U. Scherf, Nonlinear emission and recombination in conjugated polymer waveguides. Appl. Phys. Lett. **85**(2), 1124–1130 (1999)
151. W. Holzer, A. Penzkofer, S. Gong, A. Bleyer, D.D.C. Bradley, Laser action in poly(*m*-phenylenevinylene-*co*-2,5-dioctoxy-*p*-phenylenevinylene). Adv. Mater. **8**(12), 974–978 (1996)

152. M. Yan, L.J. Rothberg, F. Papadimitrakopoulos, L.E. Galvin, T.M. Miller, Spatially indirect excitons as primary photoexcitations in conjugated polymers. Phys. Rev. Lett. **72**(7), 1104–1107 (1994)
153. S.V. Frolov, Z.V. Vardeny, K. Yoshina, Cooperative and stimulated emission in poly(*p*-phenylene-vinylene) thin films and solutions. Phys. Rev. B **57**(15), 9141–9147 (1998)
154. T. Granlund, M. Theander, M. Berggren, M. Andersson, A. Ruzeckas, V. Sundström, G. Björk, M. Granström, O. Inganäs, A polythiophene microcavity laser. Chem. Phys. Lett. **288**, 879–884 (1998)
155. F. Laquai, A.K. Mishra, K. Müllen, R.H. Friend, Amplified spontaneous emission of poly(ladder-type phenylene)s—the influence of photophysical properties on ASE thresholds. Adv. Funct. Mater. **18**, 3265–3275 (2008)
156. B.K. Yap, R. Xia, M. Campoy-Quiles, M.N. Stavrinou, D.D.C. Bradley, Simultaneous optimization of charge-carrier mobility and optical gain in semiconducting polymer films. Nat. Mater. **7**, 376–380 (2088)
157. G. Wegmann, B. Schweitzer, D. Hertel, H. Giessen, M. Oestreich, U. Scherf, K. Mullen, R.F. Mahrt, The dynamics of gain-narrowing in a ladder-type π-conjugated polymer. Chem. Phys. Lett. **312**, 376–384 (1999)
158. G. Cerullo, S. Stagira, M. Nisoli, S.D. Silvestri, G. Lanzani, G. Kranzelbinder, W. Graupner, G. Leising, Excited-state dynamics of poly(para-phenylene)-type ladder polymers at high photoexcitation density. Phys. Rev. B **57**(20), 12806–12811 (1998)
159. H. Tanaka, Y. Yoshida, T. Nakao, N. Tsujimoto, A. Fujii, M. Ozaki, Photopumped laser oscillation and charge carrier mobility of composite films based on poly(3-haxylthiophene)s with different stereoregularity. Jpn. J. Appl. Phys. **45**(40), L1077–L1079 (2006)
160. T. Nguyen, I.B. Martini, J. Liu, B.J. Schwartz, Controlling interchain interactions in conjugated polymers: the effects of chain morphology on exciton-exciton annhililation and aggregation in meh-ppv films. J. Phys. Chem. B **104**, 237–255 (2000)
161. T. Nguyen, V. Doan, B.J. Schwartz, Conjugated polymer aggregates in solution: control of interchain interactions. J. Chem. Phys. **110**(8), 4068–4078 (1999)
162. M. Anni, G. Gigli, R. Cingolan, M. Zavelani-Ross, C. Gadermaier, G. Lanzani, G. Barbarella, L. Favaretto, Amplified spontaneous emission from a soluble thiophene-based oligomer. Appl. Phys. Lett. **78**(18), 2679–2681 (2001)
163. F. Laquai, P.E. Keivanidis, S. Baluschev, J. Jacob, K. Müllen, G. Wegner, Low threshold amplified spontaneous emission in thin films of poly(tetraarylindenofluorene). Appl. Phys. Lett. **87**, 261917 (2005)
164. H. Nakanotani, N. Matsumoto, H. Uchiuzou, M. Nishiyama, M. Yahiro, C. Adachi, Very low amplified spontaneous emission threshold and electroluminescence characteristics of 1,1'-diphenyl substituted fluorene derivatives. Opt. Mater. **30**, 630–636 (2007)
165. K.L. Shaklee, R.F. Leheny, Direct determination of optical gain in semiconductor crystals. Appl. Phys. Lett. **18**(11), 475–477 (1971)
166. Y. Sorek, R. Reisfeld, I. Finkelstein, S. Ruschin, Light amplification in dye-doped glass planar waveguide. Appl. Phys. Lett. **66**(10), 1169–1171 (1995)
167. G. Jordan, M. Flämmich, M. Rüther, T. Kobayashi, W.J. Blau, Y. Suzuki, T. Kaino, Light amplification at 501 nm and large nanosecond optical gain in organic dye-doped polymeric waveguides. Appl. Phys. Lett. **88**, 161114 (2006)
168. G. Heliotis, D.C. Bradley, G.A. Turnbull, I.D.W. Samuel, Light amplification and gain in polyfluorene waveguides. Appl. Phys. Lett. **81**(3), 415–417 (2002)
169. S.V. Frolov, W. Gellermann, M. Ozaki, K. Yoshino, Z.V. Vardeny, Cooperative emission in π-conjugated polymer thin films. Phys. Rev. Lett. **78**(4), 729–732 (1997)
170. G.H. Gelinck, J.M. Warman, M. Remmers, D. Neher, Narrow-band emissions from conjugated-polymer films. Chem. Phys. Lett. **265**, 320–326 (1997)
171. T. Virgili, D.G. Lidzey, M. Grell, D.D.C. Bradley, S. Stagira, M. Zavelani-Rossi, S.D. Silvestri, Influence of the orientation of liquid crystalline poly(9,9-dioctylfluorene) on its lasing properties in a planar microcavity. Appl. Phys. Lett. **80**(22), 4088–4090 (2002)
172. M. Koschorreck, R. Gehlhaar, V.G. Lyssenko, M. Swoboda, M. Hoffmann, K. Leo, Dynamics of a high-Q vertical-cavity organic laser. Appl. Phys. Lett. **87**, 181108 (2005)

173. L. Persano, P.D. Carro, E. Mele, R. Cingolani, D. Pisignano, M. Zavelani-Rossi, S. Longhi, G. Lanzani, Monolithic polymer microcavity lasers with on-top evaporated dielectric mirrors. Appl. Phys. Lett. **88**, 121110 (2006)
174. M.A. Díaz-García, F. Hide, B.J. Schwartz, M.D. McGehee, M.R. Andersson, A.J. Heeger, Plastic lasers: comparison of gain narrowing with a soluble semiconducting polymer in waveguides and microcavities. Appl. Phys. Lett. **70**(24), 3191–3193 (1997)
175. A. Schülzgen, C. Spiegelberg, M.M. Morrell, S.B. Mendes, B. Kippelen, N. Peyghambarian, Near diffraction-limited laser emission from a polymer in a high finesse planar cavity. Appl. Phys. Lett. **72**(3), 269–271 (1998)
176. X. Liu, H. Li, C. Song, Y. Liao, M. Tian, Microcavity organic laser device under electrically pumping. Opt. Lett. **34**(4), 503–505 (2009)
177. H. Kogelnik, C.V. Shank, Stimulated emission in a periodic structure. Appl. Phys. Lett. **18**(4), 152–154 (1971)
178. H. Kogelnik, C.V. Shank, Coupled-wave theory of distributed feedback lasers. J. Appl. Phys. **43**(5), 2327–2335 (1972)
179. M.D. McGehee, M.A. Diaz-Garcia, F. Hide, R. Gupta, E.K. Miller, D. Moses, A.J. Heeger, Semiconducting polymer distributed feedback lasers. Appl. Phys. Lett. **72**(13), 1536–1538 (1998)
180. G.A. Turnbull, P. Andrew, M.J. Jory, W.L. Barnes, I.D.W. Samuel, Relationship between photonic band structure and emission characteristics of a polymer distributed feedback laser. Phys. Rev. B **64**, 125122 (2001)
181. G. Heliotis, R. Xia, G.A. Turnbull, P. Andrew, W.L. Barnes, I.D.W. Samuel, D.D.C. Bradley, Emission characteristics and performance comparison of polyfluorene lasers with one- and two-dimensional distributed feedback. Adv. Funct. Mater. **14**(1), 91–97 (2004)
182. M. Reufer, S. Riechel, J.M. Lupton, J. Feldmann, U. Lemmer, D. Schneider, T. Benstem, T. Dobbertin, W. Kowalsky, A. Gombert, K. Forberich, V. Wittwer, U. Scherf, Low-threshold polymeric distributed feedback lasers with metallic contacts. Appl. Phys. Lett. **84**(17), 3262–3264 (2004)
183. C. Kallinger, M. Hilmer, A. Haugeneder, M. Perner, W. Spirkl, U. Lemmer, J. Feldmann, U. Scherf, K. Mullen, A. Gombert, V. Wittwer, A flexible conjugated polymer laser. Adv. Mater. **10**(12), 920–923 (1998)
184. V.G. Kozlov, G. Parthasarathy, P.E. Burrows, V.B. Khalfin, J. Wang, S.Y. Chou, S.R. Forrest, Structures for organic diode lasers and optical properties of organic semiconductors under intense optical and electrical excitations. IEEE J. Quantum Electron. **36**(1), 18–26 (2000)
185. T. Spehr, A. Siebert, T. Fuhrmann-Lieker, J. Salbeck, T. Rabe, T. Riedl, H.H. Johannes, W. Kowalsky, J. Wang, T. Weimann, P. Hinze, Organic solid-state ultraviolet-laser based on spiro-terphenyl. Appl. Phys. Lett. **87**, 161103 (2005)
186. D. Pisignano, L. Persano, P. Visconti, R. Cingolani, G. Gigli, G. Barbarella, L. Favaretto, Oligomer-based organic distributed feedback lasers by room-temperature nanoimprint lithography. Appl. Phys. Lett. **83**(13), 2545–2547 (2003)
187. A. Rose, Z. Zhu, C.F. Madigan, T.M. Swager, V. Bulovic, Sensitivity gains in chemosensing by lasing action in organic polymers. Nature **434**, 876–879 (2005)
188. D. Schneider, S. Hartmann, T. Benstem, T. Dobbertin, D. Heithecker, D. Metzdorf, E. Becker, T. Riedl, H. Johannes, W. Kowalsky, T. Weimann, J. Wang, P. Hinze, Wavelength-tunable organic solid-state distributed-feedback laser. Appl. Phys. B **77**, 399–402 (2003)
189. R. Xia, G. Heliotis, P.N. Stavrinou, D.D.C. Bradley, Polyfluorene distributed feedback lasers operating in the green-yellow spectral region. Appl. Phys. Lett. **87**, 031104 (2005)
190. S. Riechel, C. Kallinger, U. Lemmer, J. Feldmann, A. Gombert, V. Wittwer, U. Scherf, A nearly diffraction limited surface emitting conjugated polymer laser utilizing a two-dimensional photonic band structure. Appl. Phys. Lett. **77**(15), 2310–2312 (2000)
191. A.E. Vasdekis, G.A. Turnbull, I.D.W. Samuel, P. Andrew, W.L. Barnes, Low threshold edge emitting polymer distributed feedback laser based on a square lattice. Appl. Phys. Lett. **86**, 161102 (2005)

192. J. Stehr, J. Crewett, F. Schindler, R. Sperling, G. von Plessen, U. Lemmer, J.M. Lupton, T.A. Klar, J. Feldmann, A.W. Holleitner, M. Forster, U. Scherf, A low threshold polymer laser based on metallic nanoparticle gratings. Adv. Mater. **15**(20), 1726–1729 (2003)
193. G.A. Turnbull, P. Andrew, W.L. Barnes, I.D.W. Samuel, Operating characteristics of a semiconducting polymer laser pumped by a microchip laser. Appl. Phys. Lett. **82**(3), 313–315 (2003)
194. M. Notomi, H. Suzuki, T. Tamamura, Directional lasing oscillation of two-dimensional organic photonic crystal lasers at several photonic band gaps. Appl. Phys. Lett. **78**(10), 1325–1327 (2001)
195. N. Moll, R.F. Mahrt, C. Bauer, H. Giessen, B. Schnabel, E.B. Kley, U. Scherf, Evidence for bandedge lasing in a two-dimensional photonic bandgap polymer laser. Appl. Phys. Lett. **80**(5), 734–736 (2002)
196. G.A. Turnbull, A. Carleton, G.F. Barlow, A. Tahraouhi, T.F. Krauss, K.A. Shore, I.D.W. Samuel, Influence of grating characteristics on the operation of circular-grating distributed-feedback polymer lasers. J. Appl. Phys. **98**, 023105 (2005)
197. G. Ramos-Ortiz, C. Spiegelberg, N. Peyghambarian, B. Kippelen, Temperature dependence of the threshold for laser emission in polymer microlasers. Appl. Phys. Lett. **77**(18), 2783–2785 (2000)
198. S.V. Frolov, M. Shkunov, Z.V. Vardeny, Ring microlasers from conducting polymers. Phys. Rev. B **56**(8), 4363–4366 (1997)
199. S.V. Frolov, M. Shkunov, A. Fujii, K. Yoshino, Z.V. Vardeny, Lasing and stimulated emission in π-conjugated polymers. IEEE J. Quantum Electron. **36**(1), 2–11 (2000)
200. Y. Kawabe, C. Spiegelberg, A. Schülzgen, M.F. Nabor, B. Kippelen, E.A. Mash, P.M. Allemand, M. Kuwata-Gonokami, K. Takeda, N. Peyghambarian, Whispering-gallery-mode microring laser using a conjugated polymer. Appl. Phys. Lett. **72**(2), 141–143 (1998)
201. S.V. Frolov, A. Fujii, D. Chinn, Z.V. Vardeny, Cylindrical microlasers and light emitting devices from conducting polymers. Appl. Phys. Lett. **72**(22), 2811–2813 (1998)
202. M. Berggren, A. Dodabalapur, Z. Bao, R.E. Slusher, Solid-state droplet laser made from an organic blend with a conjugated polymer emitter. Adv. Mater. **9**(12), 968–971 (1997)
203. S.V. Frolov, A. Fujii, D. Chinn, M. Hirohata, R. Hidayat, M. Taraguchi, T. Masuda, K. Yoshino, Z.V. Vardeny, Microlasers and micro-leds from disubstituted polyacetylene. Adv. Mater. **10**(11), 869–872 (1998)
204. M. Berggren, A. Dodabalapur, R.E. Slusher, Z. Bao, Organic lasers based on förster transfer. Synth. Met. **91**, 65–68 (1997)
205. R. Gupta, M. Stevenson, A. Dogariu, M.D. McGehee, J.Y. Park, V. Srdanov, A.J. Heeger, H. Wang, Low-threshold amplified spontaneous emission in blends of conjugated polymers. Appl. Phys. Lett. **73**(24), 3492–3494 (1998)
206. M.A. Baldo, M.E. Thompson, S.R. Forrest, Phosphorescent materials for application to organic light emitting devices. Pure Appl. Chem. **71**(11), 2095–2106 (1999)
207. V.G. Kozlov, P.E. Burrows, G. Parthasarathy, S.R. Forrest, Optical properties of molecular organic semiconductor thin films under intense electrical excitation. Appl. Phys. Lett. **74**, 1057 (1999)
208. L. Ma, J. Ouyang, Y. Yang, High-speed and high-current density C_{60} diodes. Appl. Phys. Lett. **84**(23), 4786–4788 (2004)
209. N. Tessler, N.T. Harrison, R.H. Friend, High brightness polymer light-emitting diodes. Adv. Mater. **10**(1), 64–68 (1998)
210. H. Yamamoto, H. Kasajima, W. Yokayama, H. Sasabe, C. Adachi, Extremely-high-density carrier injection and transport over 12000 A/cm^2 into organic thin films. Appl. Phys. Lett. **86**(8), 083502 (2005)
211. T. Matsushima, H. Sasabe, C. Adachi, Carrier injection and transport characteristics of copper phthalocynanine thin films under low to extremely high current densities. Appl. Phys. Lett. **88**, 033508 (2006)
212. M.A. Baldo, R.J. Holmes, S.R. Forrest, Prospects for electrically pumped organic lasers. Phys. Rev. B **66**, 35321 (2002)

213. P. Andrew, G.A. Turnbull, I.D.W. Samuel, W.L. Barnes, Photonic band structure and emission characteristics of a metal-backed polymeric distributed feedback laser. Appl. Phys. Lett. **81**(6), 954–956 (2002)
214. J. Stehr, J. Crewett, F. Schindler, R. Sperling, G. von Plessen, U. Lemmer, J.M. Lupton, T.A. Klar, J. Feldmann, A.W. Holleitner, M. Forster, U. Scherf, A low threshold polymer laser based on metallic nanoparticle gratings. Adv. Mater. **15**(20), 1426–1729 (2003)
215. C. Gärtner, C. Karnutsch, U. Lemmer, C. Pflumm, The influence of annihilation processes on the threshold current density of organic laser diodes. J. Appl. Phys. **101**, 023107 (2007)
216. H. Kim, C.M. Gilmore, J.S. Horwitz, A. Piqué, H. Murata, G.P. Kushto, R. Schlaf, Z.H. Kafafi, D.B. Chrisey, Transparent conducting aluminum-doped zinc oxide thin films for organic light-emitting devices. Appl. Phys. Lett. **76**(3), 259–261 (2000)
217. H. Yamamoto, T. Oyamada, H. Sasabe, C. Adachi, Amplified spontaneous emission under optical pumping from an organic semiconductor laser structure equipped with transparent carrier injection electrodes. Appl. Phys. Lett. **84**(8), 1401–1403 (2004)
218. P. Görrn, T. Rabe, T. Riedl, W. Kowalsky, F. Galbrecht, U. Scherf, Low loss contacts for organic semiconductor lasers. Appl. Phys. Lett. **89**, 161113 (2006)
219. P. Görrn, T. Riedl, W. Kowalsky, Loss reduction in fully contacted organic laser waveguides using TE_2 modes. Appl. Phys. Lett. **91**, 041113 (2007)
220. M. Reufer, J. Feldmann, P. Rudati, A. Ruhl, D. Müller, K. Meerholz, C. Karnutsch, M. Gerken, U. Lemmer, Amplified spontaneous emission in an organic semiconductor multilayer waveguide structure including a highly conductive transparent electrode. Appl. Phys. Lett. **86**, 221102 (2005)
221. E.B. Namdas, T. Minghong, P. Ledochowitsch, S.R. Mednick, J.D. Yuen, D. Moses, A.J. Heeger, Low thresholds in polymer lasers on conductive substrates by distributed feedback nanoimprinting: progress towards electrically pumped plastic lasers. Adv. Mater. **20**, 1–4 (2008)
222. M.C. Gwinner, S. Khodabakhsh, M.H. Song, H. Schweizer, H. Giessen, H. Sirringhaus, Integration of a rib waveguide distributed feedback structure into a light-emitting polymer field-effect transistor. Adv. Funct. Mater. **19**(9), 1360–1370 (2009)
223. C. Gärtner, C. Karnutsch, C. Pflumm, U. Lemmer, Numerical device simulation of double-heterostructure organic laser diodes including current-induced absorption processes. IEEE J. Quantum Electron. **43**(11), 1006–1017 (2007)
224. V.G. Kozlov, S.R. Forrest, Lasing action in organic semiconductor thin films. Curr. Opin. Solid State Mater. Sci. **4**, 203–208 (1999)
225. N. Tessler, D.J. Pinner, V. Cleave, D.S. Thomas, G. Yahioglu, P.L. Barny, R.H. Friend, Pulsed excitation of low-mobility light-emitting diodes: implication for organic lasers. Appl. Phys. Lett. **74**, 2764–2766 (1999)
226. N. Tessler, D.J. Pinner, V. Cleave, P.K.H. Ho, R.H. Friend, G. Yahioglu, P.L. Barney, J. Gray, M. de Souza, G. Rumbles, Properties of light emitting organic materials within the context of future electrically pumped lasers. Synth. Met. **115**, 57–62 (2000)
227. N. Tessler, N.T. Harrison, D.S. Thomas, R.H. Friend, Current heating in polymer light emitting diodes. Appl. Phys. Lett. **73**(6), 732–734 (1998)
228. C. Gärtner, C. Karnutsch, J. Brückner, N. Christ, S. Uebe, U. Lemmer, P. Görrn, T. Rabe, T. Riedl, W. Kowaslsky, Loss processes in organic double-heterostructure laser diodes. Proc. SPIE **6655**, 665525 (2007)
229. J.M. Lupton, Over the rainbow. Nature **453**, 459–460 (2008)
230. D. Hertel, S. Setayesh, H. Nothofer, U. Scherf, K. Müllen, H. Bässler, Phosphorescence in conjugated poly(para-phenylene)-derivatives. Adv. Mater. **13**(1), 65–70 (2001)
231. Y. Yang, G.A. Turnbull, I. Samuel, Hybrid optoelectronics: a polymer laser pumped by a nitride light-emitting diode. Appl. Phys. Lett. **92**, 163306 (2008)
232. C. Karnutsch, M. Stroisch, M. Punke, U. Lemmer, J. Wang, T. Weimann, Laser diode-pumped organic semiconductor lasers utilizing two-dimensional photonic crystal resonators. IEEE Photonics Technol. Lett. **19**(10), 741–743 (2007)

233. D. Yokoyama, M. Moriwake, C. Adachi, Spectrally narrow emissions at cutoff wavelength from edges of optically and electrically pumped anisotropic organic films. J. Appl. Phys. **103**, 123104 (2008)
234. Y. Tian, Z. Gan, Z. Zhou, D.W. Lynch, J. Shinar, J. Kang, Q. Park, Spectrally narrowed edge emission from organic light-emitting diodes. Appl. Phys. Lett. **91**, 143504 (2007)
235. Y. Hamada, H. Kanno, T. Tsuyoshi, H. Takahashi, Red organic light-emitting diodes using an emitting assist dopant. Appl. Phys. Lett. **75**(12), 1682–1684 (1999)
236. C. Rost, D.J. Gundlach, S. Karg, W. Riess, Ambipolar organic field-effect transistor based on an organic heterostructure. J. Appl. Phys. **95**(10), 5782–5787 (2004)
237. M. Yoon, S.A. DiBenedetto, A. Facchetti, T.J. Marks, Organic thin-film transistors based on carbonyl-functionalized quaterthiophenes: high mobility n-channel semiconductors and ambipolar transport. J. Am. Chem. Soc. **127**, 1348–1349 (2005)
238. H. Yersin, *Highly Efficient Oleds with Phosphorescent Materials* (Wiley-VCH, Weinheim, 2007)
239. Y.D. Jin, Role of excitons and interfaces on optimization of electroluminescence efficiencies in organic light-emitting diodes. PhD thesis, K.U.Leuven (May 2003)
240. B.W. D'Andrade, S. Datta, S.R. Forrest, P. Djurovich, E. Polikarpov, M.E. Thompson, Relationship between the ionization and oxidation potentials of molecular organic semiconductors. Org. Electron. **6**, 11–20 (2005)
241. M. Schott, Introduction to the physics of organic electroluminescence. Acad. Sci. **1**, 381 (2000)
242. A. Kadashchuk, Y. Skryshevskii, A. Vakhnin, N. Ostapenko, V.I. Arkhipov, E.V. Emelianova, H. Bässler, Thermally stimulated photoluminescence in disordered organic materials. Phys. Rev. B **63**, 115205 (2001)
243. A. Kadashchuk, Y. Skryshevski, Y. Piryatinski, A. Vakhnin, E.V. Emelianova, V.I. Arkhipov, H. Bässler, J. Shinar, Thermally stimulated photoluminescence in poly(2,5-dioctoxy p-phenylene vinylene). J. Appl. Phys. **91**(8), 5016–5023 (2002)
244. I.D. Parker, Carrier tunneling and device characteristics in polymer light-emitting diodes. J. Appl. Phys. **75**(3), 1656–1666 (1994)
245. D.J. Gundlach, Y.Y. Lin, T.N. Jackson, S.F. Nelson, D.G. Schlom, Pentacene organic thin-film transistors—molecular ordering and mobility. IEEE Electron Device Lett. **18**, 87 (1997)
246. H. Sirringhaus, P.J. Brown, R.H. Friend, M.M. Nielsen, K. Bechgaard, B.M.W. Langeveld-Voss, A.J.H. Spiering, R.A.J. Janssen, E.W. Meijer, P. Herwig, D.M. de Leeuw, Two-dimensional charge transport in self-organized, high-mobility conjugated polymers. Nature **40**, 685–688 (1999)
247. I.D.W. Samuel, G. Rumbles, C.J. Collison, R.H. Friend, S.C. Moratti, A.B. Holmes, Picosecond time-resolved photoluminescence of ppv derivatives. Synth. Met. **84**, 497–500 (1997)
248. M. Muratsubaki, Y. Furukawa, T. Noguchi, T. Ohnishi, E. Fujiwara, H. Tada, Field-effect transistors based on poly(p-phenylenevinylene) derivatives. Chem. Lett. **33**(11), 1480–1481 (2004)
249. C. Adachi, S. Tokito, T. Tsutsui, S. Saito, Electroluminescence in organic films with three-layer structure. Jpn. J. Appl. Phys. **27**(2), L269–L271 (1988)
250. S. Schols, S. Verlaak, C. Rolin, D. Cheyns, J. Genoe, P. Heremans, An organic light-emitting diode with field-effect electron transport. Adv. Funct. Mater. **18**, 136–144 (2008)
251. S. Schols, C. McClatchey, C. Rolin, D. Bode, J. Genoe, P. Heremans, A. Facchetti, Organic light-emitting diodes with field-effect-assisted electron transport based on α,ω-diperfluorohexyl-quaterthiophene. Adv. Funct. Mater. **18**, 3645–3652 (2008)
252. S.D. Vusser, S. Stoedel, K. Myny, D. Janssen, S.D. Jonge, J. Genoe, P. Heremans, An integrated shadowmask technique for patterning small molecule organic semiconductors. Appl. Phys. Lett. **88**, 103501 (2006)
253. J. Veres, S. Ogier, S. Leeming, B. Brown, D. Cupertino, Air stable, amorphous organic films and their applications to solution processable flexible electronics. Mater. Res. Soc. Symp. Proc. **708**, BB8.7.1 (2002)

254. V. Bulovic, A. Shoustikov, M.A. Baldo, E. Bose, V.G. Kozlov, M.E. Thompson, S.R. Forrest, Bright, saturated, red-to-yellow organic light-emitting devices based on polarization-induced spectral shifts. Chem. Phys. Lett. **287**, 455–460 (1998)
255. D.J. Gundlach, K.P. Pernstich, G. Wilckens, M. Grüter, S. Haas, B. Batlogg, High mobility n-channel organic thin-film transistors and complementary inverters. J. Appl. Phys. **98**, 064502 (2005)
256. S. Tatemichi, M. Ichikawa, T. Koyama, Y. Taniguchi, High mobility n-type thin-film transistors based on N,N'-ditridecyl perylene diimide with thermal treatments. Appl. Phys. Lett. **89**, 112108 (2006)
257. T. van Woudenbergh, P.W.M. Blom, J.N. Huiberts, Electro-optical properties of a polymer light-emitting diode with an injection-limited hole contact. Appl. Phys. Lett. **82**(6), 985–987 (2003)
258. T.W. Kelley, D.V. Muyres, P.F. Baude, T.P. Smith, T.D. Jones, High performance organic thin film transistors. Mater. Res. Soc. Symp. Proc. **771**, L6.5.2 (2003)
259. L.-L. Chua, J. Zaumseil, J.-F. Chang, E.C.-W. Ou, P.K.-H. Ho, H. Sirringhaus, R.H. Friend, General observation of n-type field-effect behaviour in organic semiconductors. Nature **434**, 194–199 (2005)
260. S. Lamansky, P. Djurovich, D. Murphy, F. Abdel-Razzaq, H. Lee, C. Adachi, P.E. Burrows, S.R. Forrest, M.E. Thompson, Highly phosphorescent bis-cyclometalated iridium complexes: synthesis, photophysical characterization and use in organic light emitting diodes. J. Am. Chem. Soc. **123**(18), 4304–4312 (2001)
261. I. Sokolik, R. Priestley, A.D. Walser, R. Dorsinville, Bimolecular reactions of singlet excitons in tris(8-hydroxyquinoline) aluminum. Appl. Phys. Lett. **69**(27), 4168–4170 (1996)
262. G.E. Jabbour, Y. Kawabe, S.E. Shaheen, J.F. Wang, M.M. Morrell, B. Kippelen, N. Peyghambarian, Highly efficient and bright organic electroluminescent devices with an aluminum cathode. Appl. Phys. Lett. **71**(13), 1762–1764 (1997)
263. C. Chiang, M. Wu, D. Dai, Y. Wen, J. Wang, C. Chen, Red-emitting fluorenes as efficient emitting hosts for non-doped organic red-light-emitting diodes. Adv. Funct. Mater. **15**(2), 231–238 (2005)
264. M.A. Wolak, J. Delcamp, C.A. Landis, P.A. Lane, J. Anthony, Z. Kafafi, High-performance organic light-emitting diodes based on dioxolane-substituted pentacene derivatives. Adv. Funct. Mater. **16**, 1943–1949 (2006)
265. J.-P. Duan, P.-P. Sun, C.-H. Cheng, New iridium complexes as highly efficient orange-red emitters in organic light-emitting diodes. Adv. Mater. **15**(3), 224–228 (2003)
266. M.A. Baldo, C. Adachi, S.R. Forrest, Transient analysis of organic electrophosphorescence, II. transient analysis of triplet-triplet annihilation. Phys. Rev. B **62**(16), 10967–10977 (2000)
267. R.H. Young, C.W. Tang, A.P. Marchetti, Current-induced fluorescence quenching in organic light-emitting diodes. Appl. Phys. Lett. **80**(5), 874–876 (2002)
268. N. Nakanotani, H. Sasabe, C. Adachi, Singlet-singlet and singlet-heat annihilations in fluorescence-based organic light-emitting diodes under steady-state high current density. Appl. Phys. Lett. **86**, 213506 (2005)
269. A. Facchetti, M. Mushrush, H.E. Katz, T.J. Marks, N-type building blocks for organic electronics: a homologous family of fluorocarbon-substituted thiophene oligomers with high carrier mobility. Adv. Mater. **15**(1), 33–38 (2003)
270. S. Schols, L.V. Willigenburg, R. Müler, D. Bode, M. Debucquoy, S.D. Jonge, J. Genoe, P. Heremans, S. Lu, A. Facchetti, Influence of the contact metal on the performance of n-type carbonyl-functionalized quaterthiophene organic thin-film transistors. Appl. Phys. Lett. **93**, 263303 (2008)
271. J.A. Venables, G.D.T. Spiller, M. Hanbucken, Nucleation and growth of thin-films. Rep. Prog. Phys. **47**, 399 (1984)
272. M. Yoon, C. Kim, A. Facchetti, T.J. Marks, Gate dielectric chemical structure—organic field-effect transistor performance correlations for electron, hole, and ambipolar organic semiconductors. J. Am. Chem. Soc. **128**, 12851–12869 (2006)

273. M. Yoon, S.A. DiBenedetto, M.T. Russell, A. Facchetti, T.J. Marks, High-performance n-channel carbonyl-functionalized quaterthiophene semiconductors: thin-film transistor response and majority carrier type inversion via simple chemical protection/deprotection. Chem. Mater. **19**, 4864–4881 (2007)
274. S. Soeren, D. Janssen, S. Verlaak, J. Genoe, P. Heremans, Patterned growth of pentacene. Appl. Phys. Lett. **85**(23), 5550–5552 (2004)
275. S. Verlaak, S. Steudel, D. Janssen, P. Heremans, M.S. Deleuze, Nucleation of organic semiconductors on inert substrates. Phys. Rev. B **68**, 195409 (2003)
276. A. Dodabalapur, L. Torsi, H.E. Katz, Organic transistors: two-dimensional transport and improved electrical characteristics. Science **268**, 270–271 (1995)
277. R. Ruiz, A. Papadimitratos, A.C. Mayer, G.G. Malliaras, Thickness dependence of mobility in penatcene thin-film transistors. Adv. Mater. **17**, 1795–1798 (2005)
278. T. Muck, V. Wagner, U. Bass, M. Leufgen, J. Geurts, L.W. Molenkamp, In situ electrical characterization of DFH-4T field-effect transistors. Synth. Met. **146**, 317–320 (2004)
279. J. Grimshaw, *Organic Electrochemistry: An Introduction and a Guide*, 4th edn. (Dekker, New York, 2000). Chap. 10, Carbonyl Compounds, pp. 411–434
280. H.G.O. Becker, R. Beckert, G. Domschke, E. Fanghänel, W.D. Habicher, P. Metz, D. Pave, K. Schwetlick, *Organikum*, 21st edn. (Wiley-VCH, Weinheim, 2001). Chap. D.7 Reaktionen von Carbonylverbindungen, pp. 586–587
281. M. Hulce, T. Lavaute, Selective cycloalkanone reduction using aluminum amalgam. Tetrahedron Lett. **29**, 525–528 (1988)
282. J.C. Kotz, K.F. Purcell, *Chemistry & Chemical Reactivity*, 2nd edn. (Saunders College Publishing, Philadelphia, 1991). Chap. 21 Electrochemistry: The Chemistry of Oxidation-Reduction Reactions, pp. 851–899
283. D.R. Lide (ed.), *CRC Handbook of Chemistry and Physics*, 85th edn. (CRC Press, Boca Raton, 2004–2005), pp. 8.23–8.33
284. A.J. Bard, L.R. Faulkner, *Electrochemical Methods: Fundamentals and Applications*, 2nd edn. (Wiley, New York, 2001), p. 809
285. M.G. Mason, C.W. Tang, L.S. Hung, P. Raychaudhuri, J. Madathil, D.J. Giesen, L. Yan, Q.T. Le, Y. Gao, S.T. Lee, L.S. Liao, L.F. Cheng, W.R. Salaneck, D.A. dos Santos, J.L. Brédas, Interfacial chemistry of Alq_3 and LiF with reactive metals. J. Appl. Phys. **89**(5), 2756–2766 (2001)
286. C. Wu, G. Lee, T. Pi, Energy structures and chemical reactions at the Al/LiF/Alq_3 interfaces studied by synchrotron-radiation photoemission spectroscopy. Appl. Phys. Lett. **87**, 212108 (2005)
287. Q.T. Le, L. Yan, Y. Gao, M.G. Mason, D.J. Giesen, C.W. Tang, Photoemission study of aluminum/tris-(8-hydroxyquinoline) aluminum and aluminum/LiF/tris-(8-hydroxyquinoline) aluminum interfaces. J. Appl. Phys. **87**(1), 375–379 (2000)
288. C. Shen, A. Kahn, J. Schwartz, Chemical and electrical properties of interfaces between magnesium and aluminum and tris-(8-hydroxyquinoline) aluminum. J. Appl. Phys. **89**, 449 (2001)
289. C. Cheng, A. Kahn, J. Schwartz, Role of metal-molecule chemistry and interdiffusion on the electrical properties of an organic interface: the Al-F_{16}CuPc case. J. Appl. Phys. **90**, 6236 (2001)
290. Y. Hirose, A. Kahn, V. Aristov, P. Soukiassian, V. Bulovic, S.R. Forrest, Chemistry and electronic properties of metal-organic semiconductor interfaces: Al, Ti, In, Sn, Ag and Au on PTCDA. Phys. Rev. B **54**(19), 13748–13758 (1996)
291. M.S. Corbillon, M.A. Olazabal, M. Madariaga, Potentiometric study of aluminium-fluoride complexation equilibria and definition of the thermodynamic model. J. Sol. Chem. **37**, 567 (2008)
292. T.D. Anthopoulos, B. Singh, N. Marjanovic, N.S. Sariciftci, A.M. Ramil, H. Sitter, M. Cölle, D.M. de Leeuw, High performance n-channel organic field-effect transistors and ring oscillators based on c60 fullerene films. Appl. Phys. Lett. **89**, 213504 (2006)
293. S. Schols, A. Kadashchuk, P. Heremans, A. Helfer, U. Scherf, Triplet excitation scavenging in films of conjugated polymers. ChemPhysChem **10**, 1071–1076 (2009)

294. S. Schols, L.V. Willigenburg, S. Steudel, J. Genoe, P. Heremans, Pulsed excitation of oleds with field-effect electron transport. IEEE J. Quantum Electron. **46**(1), 62–67 (2010)
295. S. Barth, P. Müller, H. Riel, P.F. Seidler, W. Riess, H. Vestweber, H. Bässler, Electron mobility in tris(8-hydroxyquinoline)aluminum thin films determined via transient electroluminescence from single- and multilayer organic light-emitting diodes. J. Appl. Phys. **89**(7), 3711–3719 (2001)
296. T.G. Pavlopoulos, Scaling of dye lasers with improved laser dyes. Prog. Quantum Electron. **26**, 193–224 (2002)
297. J.B. Marling, D.W. Gregg, L. Wood, Chemical quenching of the triplet state in flashlamp-excited liquid organic lasers. Appl. Phys. Lett. **17**(12), 527–530 (1970)
298. H.S.R. Pappalardo, A. Lempicki, Long pulse laser emission from rhodamine 6G using cyclooctatetraene. Appl. Phys. Lett. **16**(7), 267–269 (1970)
299. H.S.R. Pappalardo, A. Lempicki, Long-pulse laser emission from rhodamine 6G. IEEE J. Quantum Electron. **6**(11), 716–725 (1970)
300. C. Rothe, S. King, A. Monkman, Long-range resonantly enhanced triplet formation in luminescent polymers doped with iridium complexes. Nat. Mater. **5**, 463–466 (2006)
301. L. Frutos, O. Castaño, M. Merchán, Theoretical determination of the singlet → singlet and singlet → triplet electronic spectra, lowest ionization potentials, and electron affinity of cyclooctatetraene. J. Phys. Chem. A **107**, 5472–5478 (2003)
302. L.M. Frutos, O. Castaño, J.L. Andrés, M. Merchán, A.U. Acuna, A theory of nonvertical triplet energy transfer in terms of accurate potential energy surfaces: the transfer reaction from π, π^* triplet donors to 1,3,5,7-cyclooctatetraene. J. Chem. Phys. **120**(3), 1208–1216 (2004)
303. L. Frutos, O. Castaño, A new algorithm for predicting triplet-triplet energy-transfer activated complex coordinate in terms of accurate potential-energy surfaces. J. Chem. Phys. **123**, 104108 (2005)
304. J. Saltiel, G.S. Hammond, Mechanisms of photochemical reactions in solution. XVII. cis-trans isomerization of the stilbenes by excitation transfer from low energy sensitizers. J. Am. Chem. Soc. **85**(16), 2515–2516 (1963)
305. A.A. Corman, I. Hamblett, M. Irvine, P. Raby, M.C. Standen, S. Yeates, Pulse radiolysis study of the cycloheptatriene triplet state: lifetime, relaxation, and nonvertical excitation. J. Am. Chem. Soc. **107**, 4404–4411 (1985)
306. A.P. Monkman, H.D. Burrows, L.J. Hartwell, L.E. Horsburgh, I. Hamblett, S. Navaratnam, Triplet energies of π-conjugated polymers. Phys. Rev. Lett. **86**(7), 1358–1361 (2001)
307. S.A. Bagnich, C. Im, H. Bässler, D. Neher, U. Scherf, Energy transfer in a ladder-type methyl-poly(para-phenylene) doped by pt(II)octaethylporphyrin. Chem. Phys. **299**, 11–16 (2004)
308. D. Hertel, H. Bässler, R. Guentner, U. Scherf, Triplet-triplet annihilation in a poly(fluorene)-derivative. J. Chem. Phys. **115**(21), 10007 (2001)
309. A. Kadashchuk, A. Vakhnin, Y. Skryshevski, V.I. Arkhipov, E.V. Emelianova, H. Bässler, Thermally stimulated luminescence in π-conjugated polymers containing fluorene and spirobifluorene units. Chem. Phys. **291**, 243–250 (2003)
310. S.A. Bagnich, H. Bässler, Origin of delayed fluorescence of a ladder-type methyl-poly(para-phenylene) doped with pt(II)octaethylporphyrin. Chem. Phys. Lett. **381**, 464–470 (2003)
311. A. Kadashchuk, S. Schols, A. Vakhnin, J. Genoe, P. Heremans, Triplet dynamics and charge carrier trapping in triplet-emitter doped conjugated polymers. Chem. Phys. **358**, 147–155 (2009)
312. M. Thompson, The evolution of organometallic complexes in organic light-emitting devices. Mater. Res. Soc. Bull. **32**, 694–701 (2007)
313. V. Cleave, G. Yahioglu, P.L. Barny, R.H. Friend, N. Tessler, Harvesting singlet and triplet energy in polymer LEDs. Adv. Mater. **11**(4), 285–288 (1999)
314. V. Cleave, G. Yahioglu, P.L. Barny, D. Hwang, A.B. Holmes, R.H. Friend, N. Tessler, Transfer processes in semiconducting polymer-porphyrin blends. Adv. Mater. **13**(1), 44–47 (2001)
315. S.A. Bagnich, H. Bässler, D. Neher, Sensitized phosphorescence of benzil-doped ladder-type methyl-poly(para-phenylene). J. Chem. Phys. **121**(18), 9178–9183 (2004)

316. F. Laquai, C. Im, A. Kadashchuk, H. Bässler, Sensitized intrinsic phosphorescence from a poly(phenylene-vinylene) derivative. Chem. Phys. Lett. **375**, 286–291 (2003)
317. Y.V. Romanovskii, H. Bässler, Phosphorescence from a ladder-type conjugated polymer in solid solutions at low temperature. Chem. Phys. Lett. **326**, 51–57 (2000)
318. A.P. Monkman, H.D. Burrows, I. Hamblett, S. Navaratnam, U. Scherf, C. Schmitt, The triplet state of the ladder-type methyl-poly(p-phenylene) as seen by pulse radiolysis-energy transfer. Chem. Phys. Lett. **327**, 111–116 (2000)
319. R.W.T. Higgins, A.P. Monkman, H. Nothofer, U. Scherf, Energy transfer to porphyrin derivative dopants in polymer light-emitting diodes. J. Appl. Phys. **91**(1), 99–105 (2002)
320. I. Tanaka, Y. Tabata, S. Tokito, Comparison of phosphorescence properties of green-emitting $Ir(ppy)_3$ and red-emitting $Btp_2Ir(acac)$. Jpn. J. Appl. Phys. **43**(12B), L1601–L1603 (2004)
321. C. Adachi, M.A. Baldo, S.R. Forrest, High-efficiency red electrophosphorescence devices. Appl. Phys. Lett. **78**(11), 1622–1624 (2001)
322. Y.V. Romanovskii, A. Gerhard, B. Schweitzer, U. Scherf, R.I. Personov, H. Bässler, Phosphorescence of π-conjugated oligomers and polymers. Phys. Rev. Lett. **84**(5), 1027–1030 (2000)
323. Y.V. Romanovskii, H. Bässler, Phosphorescence from a ladder-type conjugated polymer in solid solutions at low temperature. Chem. Phys. Lett. **326**, 51–57 (2000)
324. S.A. Patil, U. Scherf, A. Kadashchuk, New conjugated ladder polymer containing carbazole moieties. Adv. Funct. Mater. **13**(8), 609–614 (2003)
325. Y. Kawamura, J. Brooks, J.J. Brown, H. Sasabe, C. Adachi, Intermolecular interaction and a concentration-quenching mechanism of phosphorescent Ir(III) complexes in a solid film. Phys. Rev. Lett. **96**, 017404 (2006)
326. P.M. Borsenberger, D.S. Weiss, *Organic Photoreceptors for Imaging Systems* (Dekker, New York, 1993)
327. H. Kuhn, Classical aspects of energy transfer in molecular systems. J. Chem. Phys. **53**, 101–108 (1970)
328. W. Zhu, Y. Mo, M. Yuan, Y. Cao, Highly efficient electrophosphorescent devices based on conjugated polymers doped with iridium complexes. Appl. Phys. Lett. **80**(12), 2045–2047 (2002)
329. H. Liu, J. He, P. Wang, H. Xie, X. Zhang, C. Lee, B. Sun, Y. Xia, High-efficiency polymer electrophosphorescent diodes based on an Ir (III) complex. Appl. Phys. Lett. **87**, 221103 (2005)
330. V.R. Nikitenko, D. Hertel, H. Bässler, Dispersive geminate recombination in a conjugated polymer. Chem. Phys. Lett. **348**, 889–94 (2001)
331. J. Yoon, W. Lee, E.L. Thomas, Optically pumped surface-emitting lasing using self-assembled block-copolymer-distributed Bragg reflectors. Nano Lett. **6**(10), 2211–2214 (2006)
332. C. Adachi, H. Nakanotani, T. Matsushima, S. Akiyama, Y. Kawamura, Amplified spontaneous emission characteristics and low-threshold mechanism in organic solid state thin film based on styrylbenzene derivatives. Presentation at SPIE Optics and Photonics West 2006, San Diego, USA, 2006
333. M.A. Baldo, M.E. Thompson, S.R. Forrest, High-efficiency fluorescent organic light-emitting devices using a phosphorescent sensitizer. Nature **403**, 750–753 (2000)
334. J. Nagle, S. Hersee, M. Krakowski, T. Weil, C. Weisbuch, Threshold current of single quantum well lasers: the role of the confining layers. Appl. Phys. Lett. **49**(20), 1325–1327 (1986)
335. P.L. Derry, A. Yariv, K.Y. Lau, N. Bar-Chaim, K. Lee, J. Rosenberg, Ultralow-threshold graded-index separate-confinement single quantum well buried heterostructure (Al,Ga)As lasers with high reflectivity coatings. Appl. Phys. Lett. **50**(25), 1773–1775 (1987)

MIX
Papier aus verantwortungsvollen Quellen
Paper from responsible sources
FSC® C105338

If you have any concerns about our products,
you can contact us on
ProductSafety@springernature.com

In case Publisher is established outside the EU,
the EU authorized representative is:
Springer Nature Customer Service Center GmbH
Europaplatz 3, 69115 Heidelberg, Germany

Printed by Libri Plureos GmbH
in Hamburg, Germany